Table of Contents

The Purpose of This Book 11

Introduction to Calculus 12

Limits and Continuity 13

    Defining Limits and Using Limit Notation 13

        Practice 1 17

    Finding Limits from Graphs and Tables 18

        GRAPHING CALCULATOR GREATEST INTEGER 23

        Practice 2 24

    Finding Limits Algebraically 26

        Direct Substitution 27

        Vertical Asymptote 27

        Using Factoring to Find Limits 29

        Using Conjugates to Find Limits 29

        Simplifying Complex Fractions to Find Limits 30

        Using Trig Identities to Solve Limits 31

        Practice 3 32

    Squeeze Theorem (aka Sandwich Theorem) 34

Practice 4 37

Continuity 38

Types of Discontinuity 40

Vertical Asymptotes and Limits 43

Horizontal Asymptotes and Limits 44

Practice 5 45

Intermediate Value Theorem 48

Practice 6 50

Differentiation: Definition and Basic Rules 51

Average Rate of Change vs. Instantaneous Rate of Change 52

Estimating the Derivative at a Point 53

Practice 7 55

Definition of the Derivative 56

Practice 8 62

Differentiability 65

Practice 9 69

Basic Derivative Rules 71

Power Rule for Derivatives 72

Practice 10 74

Product Rule for Derivatives 76

    Practice 11 77

Quotient Rule for Derivatives 78

    Practice 12 79

Derivatives of Exponential and Logarithmic Functions 80

    Practice 13 82

Derivatives of Trigonometric Functions 82

    Practice 14 84

Differentiation: Composite, Implicit, and Inverse 88

    Chain Rule 88

        Practice 15 91

    Implicit Differentiation 93

        Practice 16 96

    Derivatives of Inverse Functions 99

        Practice 17 101

    Derivatives of Inverse Trigonometric Functions 104

        Practice 18 105

    Derivatives of Higher Order 108

        Practice 19 110

- Contextual Applications of Differentiation 113
  - Interpreting the Meaning of the Derivative in Context 113
    - GRAPHING CALCULATOR: DERIVATIVE AT A POINT 115
      - Practice 20 116
  - Motion Problems 118
    - Interpreting Motion Graphs 120
    - Practice 21 126
    - Motion Problems Using Calculus 130
    - Practice 22 133
  - Related Rates 136
    - Practice 23 143
  - Local Linearity 149
    - Practice 24 151
  - Indeterminate Forms and L'Hôpital's Rule 154
    - Practice 25 157
- Applying Derivatives to Analyze Functions 161
  - Intermediate Value Theorem 161
    - Practice 26 163
  - Mean Value Theorem 164

Practice 27 166

Extrema 169

Extreme Value Theorem 169

Practice 28 171

Intervals of Increasing and Decreasing 174

Practice 29 177

First Derivative Test 181

Concavity and Inflection Points 181

Second Derivative Test 182

Curve Sketching 183

Practice 30 188

Relationship among $f x$, $f'\ x$, and $f'\ '\ x$ 194

Practice 31 197

Optimization Problems 202

Practice 32 205

Integration and Accumulation of Change 214

Accumulation of Change 214

Practice 33 217

Approximating Area with Riemann Sums 220

Practice 34 222

GRAPHING CALCULATOR: DEFINITE INTEGRAL 224

Approximating Area with Trapezoidal Rule 225

Practice 35 228

Riemann Sums and Definite Integrals 230

Practice 36 232

Properties of Definite Integrals 235

Practice 37 235

Fundamental Theorem of Calculus 237

Practice 38 238

Finding Antiderivatives and Indefinite Integrals 241

Reverse Power Rule 241

Practice 39 243

Integrals Using u-Substitution 247

Practice 40 249

Integrals Involving ex and $1x$ 253

Practice 41 254

Integrals Involving Trigonometric Functions 257

Practice 42 259

Definite Integrals 264

　　Practice 43 265

Integrals Involving Inverse Trigonometric Functions 267

　　Practice 44 268

SOLUTIONS 271

　　Solutions Practice 1 271

　　Solutions Practice 2 271

　　Solutions Practice 3 271

　　Solutions Practice 4 271

　　Solutions Practice 5 272

　　Solutions Practice 6 272

　　Solutions Practice 7 272

　　Solutions Practice 8 273

　　Solutions Practice 9 273

　　Solutions Practice 10 274

　　Solutions Practice 11 274

　　Solutions Practice 12 275

　　Solutions Practice 13 275

　　Solutions Practice 14 276

Solutions Practice 15 276

Solutions Practice 16 277

Solutions Practice 17 278

Solutions Practice 18 278

Solutions Practice 19 279

Solutions Practice 20 280

Solutions Practice 21 280

Solutions Practice 22 282

Solutions Practice 23 283

Solutions Practice 24 283

Solutions Practice 25 284

Solutions Practice 26 284

Solutions Practice 27 285

Solutions Practice 28 285

Solutions Practice 29 286

Solutions Practice 30 288

Solutions Practice 31 293

Solutions Practice 32 294

Solutions Practice 33 296

Solutions Practice 34 296

Solutions Practice 35 297

Solutions Practice 36 297

Solutions Practice 37 298

Solutions Practice 38 299

Solutions Practice 39 299

Solutions Practice 40 300

Solutions Practice 41 300

Solutions Practice 42 301

Solutions Practice 43 301

Solutions Practice 44 301

APPENDIX I: TABLES 302

Table 1: Calculator Quick Reference 302

Table 2: Properties of Limits 304

Table 3: Properties of Derivatives 305

Table 4: Derivative Formulas 306

Table 5: Common Formulas for Related Rates 307

Table 6: Properties of Definite Integrals 308

Table 7: Integral Formulas 309

References 310

GRAPH PAPER 311

GRAPH PAPER TRIG 319

GRAPH PAPER FULL 321

The Purpose of This Book

The purpose of this book is to provide a basic understanding of Calculus at the advanced high school or beginning of college. This past year has been my first year teaching AP Calculus AB. As usual, I was supplied with a textbook for my students and myself. I tried using the textbook to see what matched with the current AP Calculus curriculum. What I discovered was the textbook that was provided for me and my students wasn't really helpful. The textbook was not reader friendly; it did not have a lot of relevant examples worked out; it did not have any conceptual and critical thinking questions; and it did not follow the AP curriculum. I had to make a lot of notes, classwork and homework sets on my own. I also used and assigned exercises from Khan Academy which follows the curriculum quite well. So hopefully if you are taking Calculus for the first time whether it is AP Calculus in high school or Calc 1 in college you will find this book helpful. This book will go through most of what is expected in Calculus 1 or AP Calc AB.

This book is an expanded form of my lecture notes and includes extra explanations, examples, and practice. If you get stuck with the practice or just want to check your answers, then check with BOB. Solutions to practice sets are in the Back Of the Book. Throughout the book you will also find GRAPHING CALCULATOR HELP sections which will guide you through using a TI-84 series graphing calculator.

Calculus is a notoriously difficult class for at least three reasons: 1) It is a different type of mathematics than students have seen before, 2) It relies on all the mathematics from Algebra I, Geometry, Algebra II, Trigonometry and Precalc classes previously taken, 3) There is a lot of material to cover in a short time. But the basic concepts of calculus are not that difficult. In order to overcome these difficulties, I will attempt to make the material as intuitive and visual as possible, remind/reteach you of all the previously material you probably learned before but have now forgotten, and break the material into manageable chunks with just enough practice for each concept.

Introduction to Calculus

Calculus is a branch of mathematics dealing with change. Specifically, we may have a function or rule which tells us how one variable will change when a change is made in another variable. For example, $f(x) = x^2$ is a function that tells us that $f(x)$ or y is equal to the square of x. When x is 1, y is 1. When x is 2, y is 4. When x is 3, y is 9. For small changes in x, y can change quite a bit. We may wonder how fast does y change. Well it depends on the x and it's difficult to find an exact answer using just algebra. Calculus can help us find the answer.

There are two main branches of calculus, differential calculus and integral calculus. Differential calculus is used to divide a function into the smallest imaginable part to give information about how the function changes at a particular point while integral joins up these small pieces to determine what is accumulated over time. Visually, differentiation gives us the slope at any point while integration tells us about the area under the function. In order to do both of these processes, we need to think of very tiny, infinitesimal pieces of a function. Infinitesimal just means the pieces get closer and closer to zero.

Limits and Continuity

Defining Limits and Using Limit Notation

Limits are fundamental to the understanding of calculus because we want to either get as close to infinity as possible or get as close to an infinitesimal as possible. Infinity is not really a number but consider to be bigger than any possible real number, while an infinitesimal is immeasurably small number as close to zero as we can get. A limit is the value a function's output (usually y value) approaches as the input (usually x) approaches some other value. Mathematically this is written as:

$$\lim_{x \to c} f(x) = L$$

which you read as "the limit of f as x approaches c is equal to L."

There is a formal definition of limits called the "epsilon-delta" definition:

$\lim_{x \to c} f(x) = L$ if for every $\varepsilon > 0$, there exists a $\delta > 0$ such that, for all $x \in D$, if $0 < |x - c| < \delta$, then $|f(x) - L| < \varepsilon$.

This just means that if you want to get really close to the limit of the function, you just have to pick a value close enough to c from either side in order to get it to work.

Ex. 1 Find the limit: $\lim\limits_{x \to 4} x - 1$

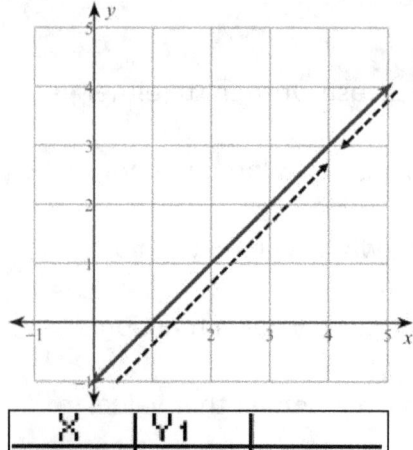

Solution: We want to find the limit of the function $f(x) = x - 1$ as x gets closer and closer to 4. Technically, we can't just assume we can plug in 4 into the function. We can explore limits algebraically, graphically, and from a table. From the graph, you can see that when x approaches 4 from either the left side or the right side, the y value gets closer and closer to 3. We can see this on the graph as we travel along the line. We can also see this on the table of values. In this case, we can find the limit by direct substitution as well, $f(4) = 4 - 1 = 3$, but for some functions this is not always the case. *We can conclude that* $\lim\limits_{x \to 4} x - 1 = 3.$

Ex. 2 Find the limit: $\lim\limits_{x \to 4} \dfrac{x^2 - 5x + 4}{x - 4}$

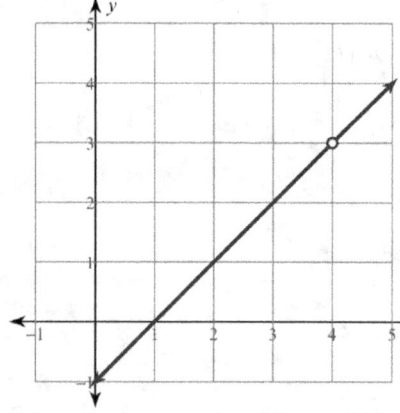

Solution: Here in this example, we want to find the limit of the function $f(x) = \dfrac{x^2 - 5x + 4}{x - 4}$ as x gets closer and closer to 4. If we try to just substitute 4 in for x we get $\dfrac{0}{0}$, which is undefined, or more specifically it is an indeterminate form. When you graph the function, it looks like a line from the previous example, with one important exception. There is an open circle at the point (4,3).

This is because the function is not defined at x=4. But as x approaches 4 from either the left side or the right side, the y value still gets closer and closer to 3. We can see this on the graph as we travel along the line. We can also see this on the table of values. If we simplify this rational expression by factoring the numerator, we get

$\frac{x^2-5x+4}{x-4} = \frac{(x-4)(x-1)}{x-4} = x - 1, x \neq 4$. So, this function is equivalent to the previous at all points except at x = 4. Therefore, $\lim_{x \to 4} \frac{x^2-5x+4}{x-4} = 3$.

Ex. 3 Find the limit: $\lim_{x \to 1} x^3 - 3x^2 - 2$

Solution: When trying to find a limit you should always first try direct substitution. $(1)^3 - 3(1)^2 - 2 = -4$. Therefore, $\lim_{x \to 1} x^3 - 3x^2 - 2 = -4$.

Properties of Limits

Let b and c be real numbers, let n be a positive integer, and let f and g be functions with the following limits: $\lim_{x \to c} f(x) = L$ and $\lim_{x \to c} g(x) = K$

1.	Constant: $\lim_{x \to c} b = b$
2.	Limit of x: $\lim_{x \to c} x = c$
3.	Scalar Multiple: $\lim_{x \to c}[bf(x)] = bL$
4.	Sum or Difference: $\lim_{x \to c}[f(x) \pm g(x)] = L \pm K$
5.	Product: $\lim_{x \to c}[f(x)g(x)] = LK$
6.	Quotient: $\lim_{x \to c}\left[\frac{f(x)}{g(x)}\right] = \frac{L}{K}$, as long as $K \neq 0$
7.	Power: $\lim_{x \to c}[f(x)]^n = L^n$
8.	Root: $\lim_{x \to c} \sqrt[n]{f(x)} = \sqrt[n]{L}$

<u>Ex. 4</u> Given that $\lim_{x \to 2} f(x) = 12$ and $\lim_{x \to 2} g(x) = 4$, find the limit of $\lim_{x \to 2}[f(x) - 2g(x) + 3]$

Solution: We can use properties of limits to break up this single unit into manageable and knowable parts:

$$\lim_{x \to 2}[f(x) - 2g(x) + 3] = \lim_{x \to 2} f(x) - 2\lim_{x \to 2} g(x) + \lim_{x \to 2} 3 = 12 - 2(4) + 3 = 7$$

We used the constant rule, the scalar multiple rule and the sum/difference rules. Then we can find the limits of each part separately and find our answer.

Practice 1

Evaluate each limit.

1) $\lim_{x \to 2} (2x + 3)$

2) $\lim_{x \to -2} (-x^2 + 2x + 1)$

3) $\lim_{x \to 2} \sqrt{x + 5}$

4) $\lim_{x \to -2} \dfrac{x^2 + 4x - 5}{x + 5}$

5) $\lim_{x \to -\frac{\pi}{2}} \sin(x)$

6) $\lim_{x \to \pi} 4$

7) $\lim_{x \to \frac{3\pi}{4}} \cos^2(x)$

8) $\lim_{x \to \frac{\pi}{3}} \sqrt{4\sin^2(x) + 1}$

Given that $\lim_{x \to 4} f(x) = 6$, $\lim_{x \to 4} g(x) = 3$, $\lim_{x \to 4} h(x) = 2$, determine the following:

9) Find $\lim_{x \to 4} 2f(x)$

10) Find $\lim_{x \to 4} (f(x) + g(x))$

11) Find $\lim_{x \to 4} f(x) \cdot g(x)$

12) Find $\lim_{x \to 4} \dfrac{f(x)}{h(x)}$

13) Find $\lim_{x \to 4} (h(x))^2$

14) Find $\lim_{x \to 4} \dfrac{2g(x) + f(x)}{4h(x) - 2}$

Finding Limits from Graphs and Tables

Ex. 1

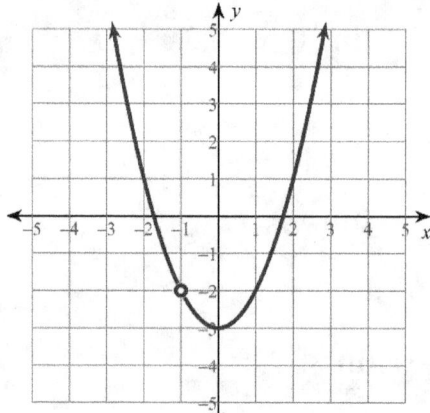

The graph of $f(x)$ is shown above. Find $\lim\limits_{x \to -1} f(x)$

Solution: For this graph, we can't just evaluate the limit by substituting -1 in for x, because the function is not defined at x = -1, there is an open circle there. When we get close to x=-1, from either the left side or the right side we see that the y values get closer and closer to -2. Therefore $\lim\limits_{x \to -1} f(x) = -2$.

Ex. 2

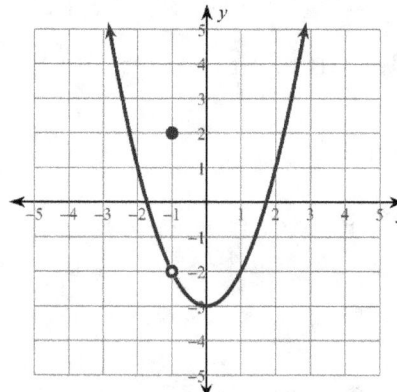

The graph of $f(x)$ is shown above. Find $\lim\limits_{x \to -1} f(x)$

Solution: This graph is like the graph above with an open circle, but now the graph also has a point defined at (-1, 2). But still as we approach x = -1, from either side we get closer to -2. The actual value $f(-1) = 2$ does not matter when we are finding the limit. All that matters is what happens when we get close to x = -1. Therefore, we still have $\lim\limits_{x \to -1} f(x) = -2$.

Ex. 3

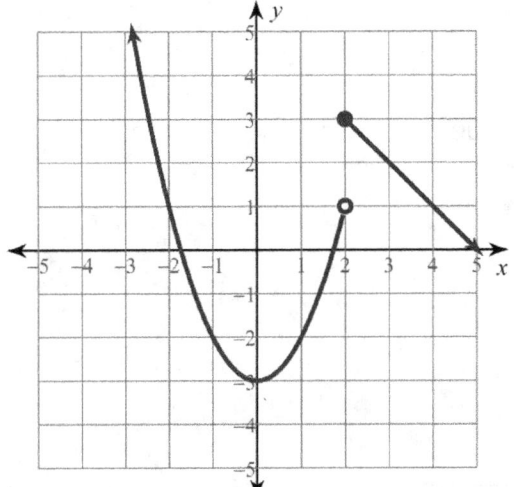

The graph of $f(x)$ is shown above. Find $\lim_{x \to 2} f(x)$

Solution: For this problem when we approach x = 2 from the left we get closer and closer to 1, but as we approach x = 2 from the right side we get closer to closer to 3. For there to be a limit we must get closer to the same value from both sides. Since this is not the case here, we ultimately have $\lim_{x \to 2} f(x)$ does not exist, sometimes abbreviated DNE. That's right, the limit does not exist if we don't reach the same finite value from both the left side and the right side.

Ex. 4

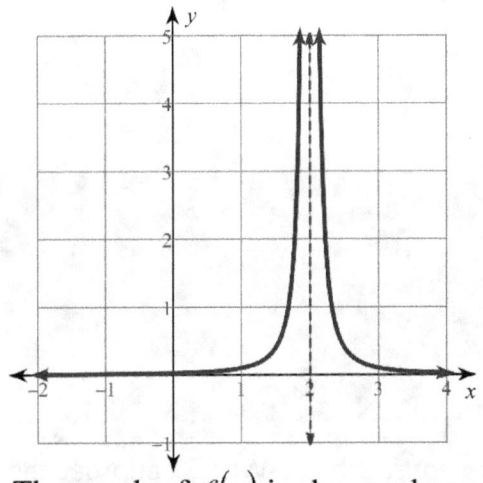

The graph of $f(x)$ is shown above. Find $\lim_{x \to 2} f(x)$

Solution: Now this graph is a little different. What happens when we get close to x = 2 from either side? It seems to shoot up towards infinity. There is a vertical asymptote at x = 2, so as we get close to x = 2, y just gets bigger and bigger. In this case it goes up on both sides. Therefore, we have $\lim_{x \to 2} f(x) = \infty$, since infinity is not really a number we may also say that $\lim_{x \to 2} f(x)$ does not exist.

Ex. 5 Using the table, what is a reasonable estimate for $\lim\limits_{x \to 2} f(x)$

x	1.5	1.9	1.99	2.01	2.1	2.5
$f(x)$	4.4	4.12	4.01	3.99	3.94	3.6

Solution: When x approaches 2 from either side, what y value does it seem we approach? If you thought 4, you are correct. $\lim\limits_{x \to 2} f(x) = 4$

Ex. 6 Using the table, what is a reasonable estimate for $\lim\limits_{x \to 4} f(x)$

x	3.5	3.9	3.99	4	4.01	4.1	4.5
y	5.5	5.9	5.99	8	6.01	6.1	6.5

Solution: When x approaches 4 from either side, it seems we get closer to 6. However, if we evaluate the function at x = 4, we get the value of y = 8. What is the limit this time? If you thought 6, you are also correct. $\lim\limits_{x \to 4} f(x) = 6$. Just like in the graphs, it matters what we approach on both sides, not on what the function actually is at that particular x-value.

Ex. 7 Using the table, what is a reasonable estimate for $\lim\limits_{x \to 3} f(x)$

x	2.6	2.9	2.99	3.01	3.1	3.6
y	−4.5	−4.9	−4.99	−7.01	−7.1	−7.5

Solution: When x approaches 3 from the left side, y seems to approach −5. When x approaches 3 from the right side, y seems to approach −7. Since we approach different values from each side, we conclude $\lim\limits_{x \to 3} f(x)$ does not exist.

Right-handed Limit

We say $\lim_{x \to c^+} f(x) = L$ provided that we can make $f(x)$ as close to L as we want for all x sufficiently close to c with $x > c$ without actually letting x be a.

Left-handed Limit

We say $\lim_{x \to c^-} f(x) = L$ provided that we can make $f(x)$ as close to L as we want for all x sufficiently close to c with $x < c$ without actually letting x be a.

Theorem: If $\lim_{x \to c^+} f(x) = \lim_{x \to c^-} f(x) = L$, then $\lim_{x \to c} f(x) = L$

$$\lim_{x \to 0^-} f(x), \; f(x) = \begin{cases} -x + 6, & x < 0 \\ -2x + 9, & x \geq 0 \end{cases}$$

Ex. 8 Find the limit

Solution: We want to find the limit as we approach 0 from the left side. Since this is a piecewise function, we will use how it is defined when x < 0. We can find the limit by direct substitution: $-(0) + 6 = 6$. Therefore $\lim_{x \to 0^-} f(x) = 6,$.

Ex. 9 Find the limit $\lim_{x \to 0^+} f(x), \; f(x) = \begin{cases} -x + 6, & x < 0 \\ -2x + 9, & x \geq 0 \end{cases}$

Solution: We want to find the limit as we approach 0 from the right side. In this case, we will use how it is defined when $x \geq 0$. We can find the limit by direct substitution: $-2(0) + 9 = 9$. Here we have, $\lim_{x \to 0^+} f(x) = 9$.

Ex. 10 Find the limit $\lim\limits_{x \to 3^-} \lfloor x \rfloor$

Solution: Here we want to find the limit of the greatest integer function as x approaches 3 from the left side. The **greatest integer function** is self-explanatory, it gives the largest integer less than or equal to the input. The greatest integer is sometimes called the **floor function**, or **step function**[1], and it denoted by $\lfloor x \rfloor$, or sometimes by $[x]$, $[\![x]\!]$.

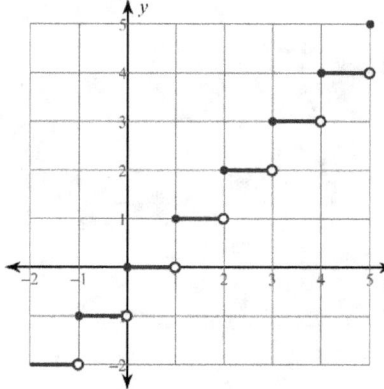

The graph shows the function's step behavior. When we approach 3 from the left, we can see that the y value will be 2. The greatest integer less than or equal to 2.9999 is 2 or we can write $\lfloor 2.9999 \rfloor = 2$. Therefore, the limit $\lim\limits_{x \to 3^-} \lfloor x \rfloor = 2$. By the way, $\lim\limits_{x \to 3^+} \lfloor x \rfloor = 3$.

---

[1] Step function is somewhat ambiguous, because there is another step function called the ceiling function, or least integer function denoted by $\lceil x \rceil$ (Wikipedia contributors, 2020)

GRAPHING CALCULATOR GREATEST INTEGER

To use the greatest integer function on the graphing calculator, press the [MATH] button. The greatest integer function is denoted by `int(`

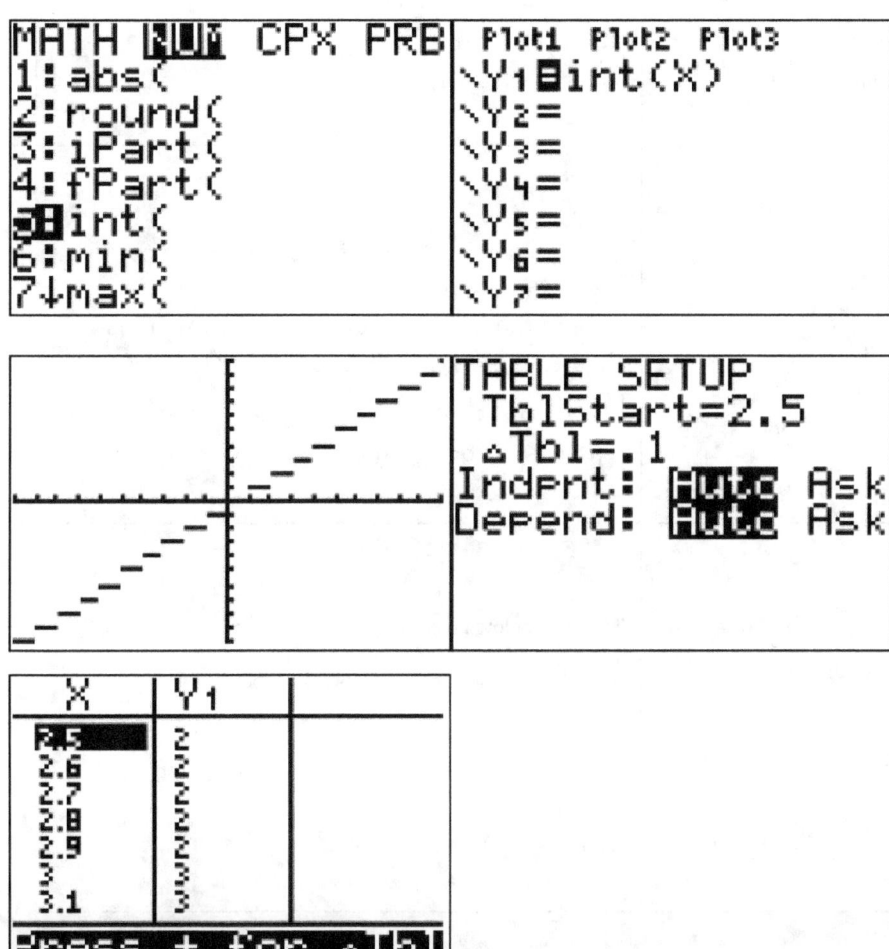

Practice 2

1.

The table below gives selected values for a continuous function f. Based on the data in the table, which of the following is the best approximation for $\lim_{x \to 4} f(x)$?

x	3.9	3.99	3.999	4.001	4.01	4.1
y	12.41	12.94	12.994	13.006	13.06	13.61

2.

The table below gives selected values for a continuous function f. Based on the data in the table, which of the following is the best approximation for $\lim_{x \to 2} f(x)$?

x	1.9	1.99	1.999	2.001	2.01	2.1
y	−5.1	−5.01	−5.001	−5.999	−5.99	−5.9

3.

The table below gives selected values for a continuous function f. Based on the data in the table, which of the following is the best approximation for $\lim_{x \to -5} f(x)$?

x	−5.1	−5.01	−5.001	−4.999	−4.991	−4.9
y	6.1	6.01	6.001	5.999	5.99	5.9

4.

The table below gives selected values for a continuous function f. Based on the data in the table, which of the following is the best approximation for $\lim_{x \to 2^+} f(x)$?

x	1.9	1.99	1.999	2.001	2.01	2.1
y	9	97	978	8.006	8.06	8.61

5.

The table below gives selected values for a continuous function f. Based on the data in the table, which of the following is the best approximation for $\lim_{x \to 2^-} f(x)$?

x	1.9	1.99	1.999	2.001	2.01	2.1
y	9	97	978	8.006	8.06	8.61

6.

The table below gives values of a function f at selected values of x. Which of the following conclusions is supported by the data in the table?

x	4.9	4.99	4.999	4.9999	5.0001	5.001	5.01	5.1
y	6.9	6.99	6.999	6.9999	−740	260	−60	20

A) $\lim_{x \to 5} f(x) = 7$ B) $\lim_{x \to 5^+} f(x) = 7$

C) $\lim_{x \to 5^-} f(x) = 7$ D) $\lim_{x \to 7^-} f(x) = 5$

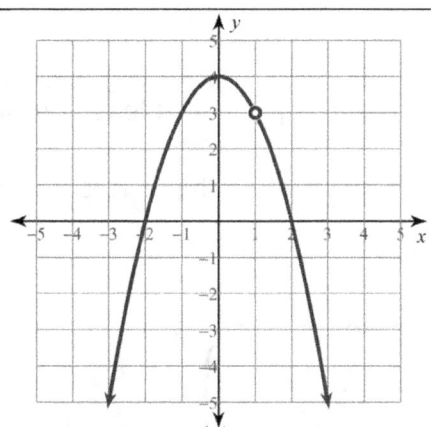

7. The graph of $f(x)$ is shown above. Find $\lim_{x \to 1} f(x)$

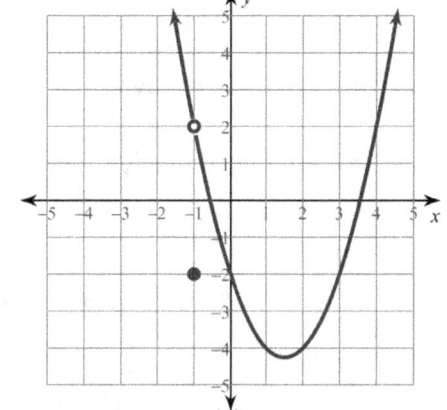

8. The graph of $f(x)$ is shown above. Find $\lim_{x \to -1} f(x)$

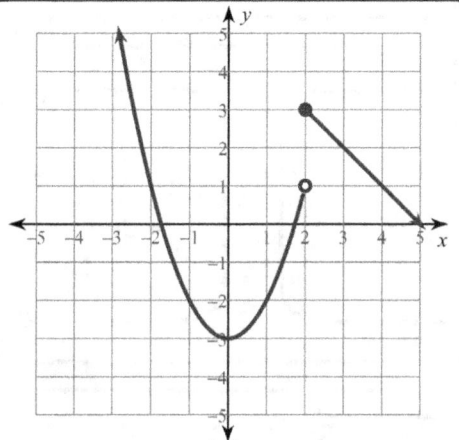

9. The graph of $f(x)$ is shown above. Find $\lim_{x \to 2} f(x)$

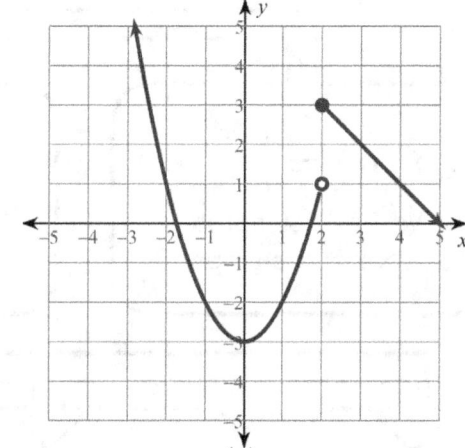

10. The graph of $f(x)$ is shown above. Find $\lim_{x \to 2^-} f(x)$

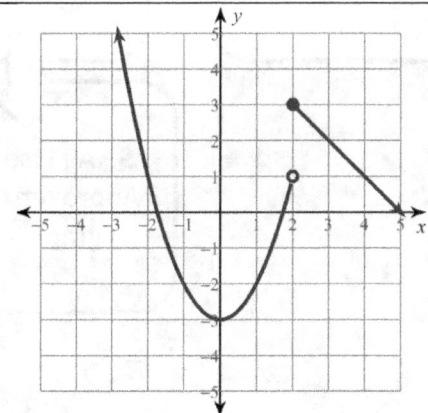

11. The graph of $f(x)$ is shown above. Find $\lim_{x \to 2^+} f(x)$

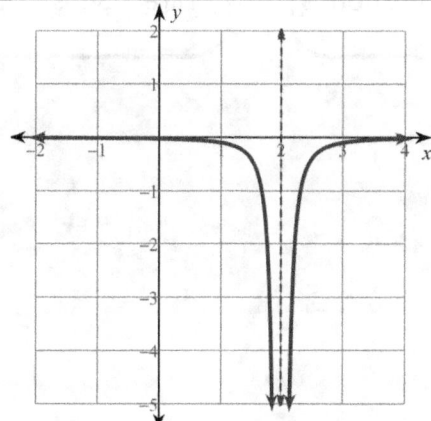

12. The graph of $f(x)$ is shown above. Find $\lim_{x \to 2} f(x)$

Finding Limits Algebraically

You may not always have the table, graph or even a calculator available to you, so how do you solve a limit algebraically? Below is a flowchart[2] demonstrating the process for finding limits algebraically.

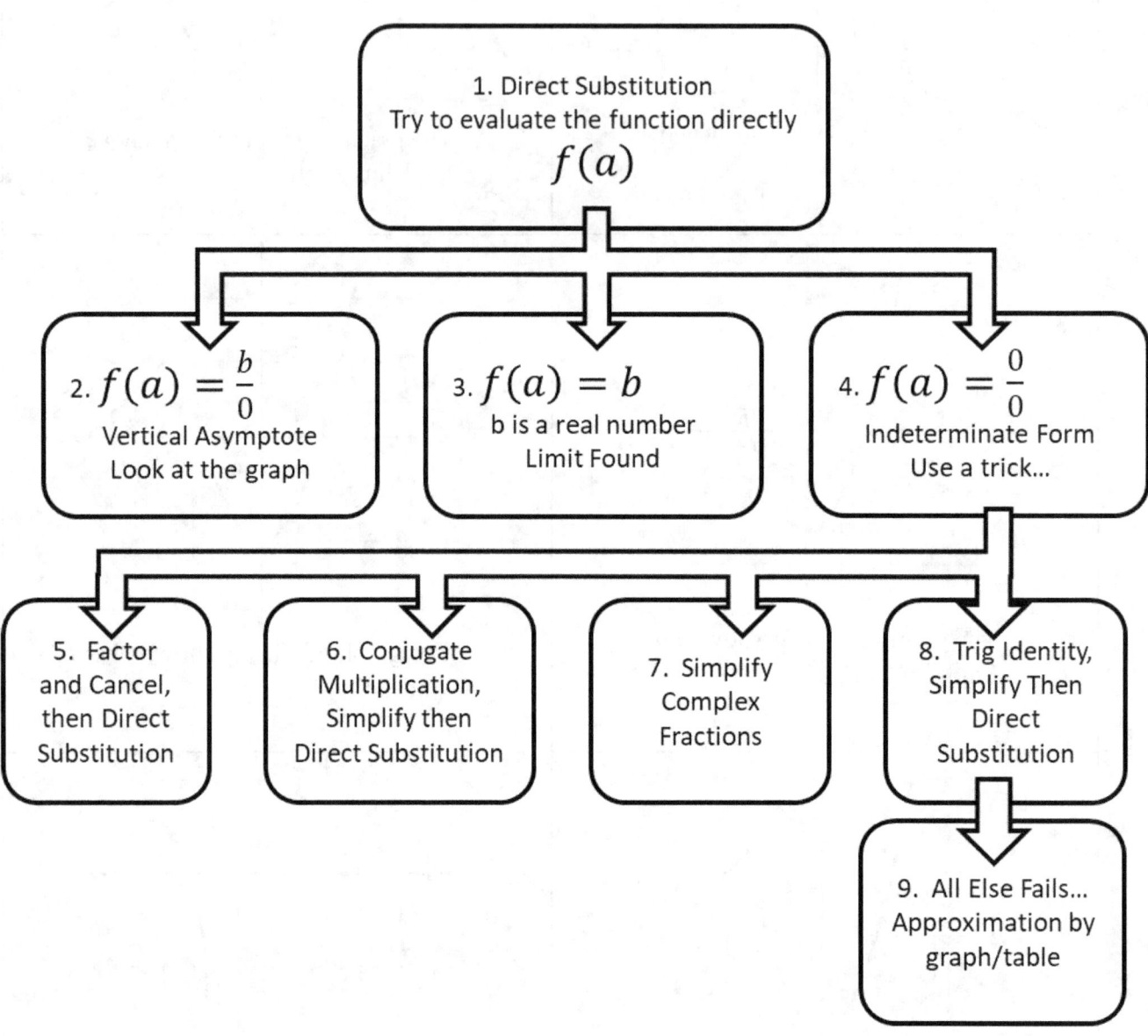

---

[2] This flowchart is modified from one given on Khan Academy (Khan Academy, 2020)

Direct Substitution

Ex. 1 Find the limit $\lim\limits_{x \to 3} \frac{x^2+2x-3}{x-1}$

Solution: So, in this limit you should first try direct substitution. $f(3) = \frac{3^2+2(3)-3}{3-1} = 6$.

Therefore, $\lim\limits_{x \to 3} \frac{x^2+2x-3}{x-1} = 6$

Vertical Asymptote

Ex. 2 Find the limit $\lim\limits_{x \to 1} \frac{2}{x-1}$

Solution: Here when we do direct substitution, we get $\lim\limits_{x \to 1} \frac{2}{x-1}$ which has the form[3] of $\frac{2}{0}$. Since we have the form $\frac{b}{0}$, we suspect that there is probably a vertical asymptote. Graphing is the easiest way to determine if it has an asymptote, but if graphing is not available, just imagine plugging in numbers close to 1 on both sides. There is no need to be exact, we just need to keep track of relatively big vs. relatively small and positive vs. negative numbers. For example, $f(.999) = \frac{2}{-.001}$ which means we have 2 on top but we are dividing by a very small[4], negative number in the denominator. The closer we get to 1 from the left, the smaller the bottom becomes.

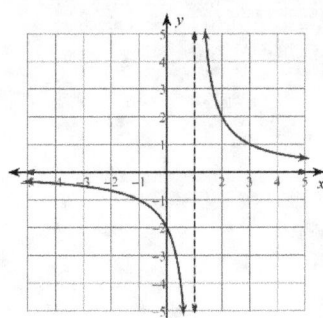

A positive number divided by a really, really small negative number, gives a really, really BIG negative number. So, as we get closer to 1 from the left, this plummets down towards negative

---

[3] We usually don't write that $f(1) = \frac{1+1}{1^2-2(1)+1}$ equals $\frac{2}{0}$ since it is undefined.

[4] Small in absolute value

infinity. Now we will approach from the right side, $f(1.001) = \frac{2}{.001}$ which means we have 2 on top but this time we are dividing by a very small, positive number in the denominator. Therefore, this side shoots up to positive infinity. We can also confirm our findings by the graph. We have $\lim_{x \to 1^-} \frac{2}{x-1} = -\infty$ and $\lim_{x \to 1^+} \frac{2}{x-1} = \infty$. Therefore, $\lim_{x \to 1} \frac{2}{x-1}$ does not exist.

<u>Ex. 3</u> Find the limit $\lim_{x \to 1} \frac{x+1}{x^2-2x+1}$

<u>Solution:</u> Here when we try direct substitution, we get $f(1) = \frac{1+1}{1^2-2(1)+1}$ which has the form of $\frac{2}{0}$. From the left side, $f(.999) = \frac{1+.999}{.999^2-2(.999)+1}$ which means we have 1.999 on top but we are dividing by a very small, positive number in the denominator so, as we get closer to 1, this shoots up towards infinity. Now approach from the right side, $f(1.001) = \frac{1+1.001}{1.001^2-2(1.001)+1}$ which means we have 2.001 on top but again we are dividing by a very small, positive number in the denominator. Therefore, this side also goes to infinity.

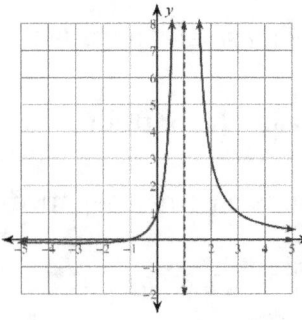

From the graph we can confirm that both the left-hand and right-hand limits as we approach 1, are infinity. Therefore, $\lim_{x \to 1} \frac{x+1}{x^2-2x+1} = \infty$ or *we can say that the limit does not exist.*

Using Factoring to Find Limits

Ex. 4 Find the limit $\lim_{x \to 1} \frac{x^2-5x+4}{x-1}$

Solution: Here when we try direct substitution, we get $\frac{0}{0}$, an indeterminate form. This means we should try some algebraic trick to change it to a different form. Since this is a rational function, we try to factor the numerator and denominator, cancelling out common factors.

$\text{Lim}_{x \to 1} \frac{x^2-5x+4}{x-1} = \lim_{x \to 1} \frac{(x-1)(x-4)}{x-1} = \lim_{x \to 1} x - 4 = 1 - 4 = -3$. Therefore, $\lim_{x \to 1} \frac{x^2-5x+4}{x-1} = -3$.

Ex. 5 Find the limit $\lim_{x \to 2} \frac{x^2-6x+8}{x^3-8}$

Solution: Here's another indeterminate form and factoring is the key. $\text{Lim}_{x \to 2} \frac{x^2-6x+8}{x^3-8} =$

$\lim_{x \to 2} \frac{(x-2)(x-4)}{(x-2)(x^2+2x+4)} = \lim_{x \to 2} \frac{x-4}{x^2+2x+4} = \frac{2-4}{2^2+2(2)+4} = \frac{-2}{12} = -\frac{1}{6}$. Therefore, $\lim_{x \to 2} \frac{x^2-6x+8}{x^3-8} = -\frac{1}{6}$

Using Conjugates to Find Limits

Ex. 6 Find the limit $\lim_{x \to 4} \frac{\sqrt{x}-2}{x-4}$

Solution: Direct substitution yields $\frac{0}{0}$ and factoring doesn't really work well[5]. When radicals appear in a limit problem with an indeterminate form, the trick is to multiply both the numerator and denominator by the **conjugate**[6] of the radical expression. The conjugate means

---

[5] It is possible to factor using radicals, ex. $x - 4 = (\sqrt{x} + 2)(\sqrt{x} - 2)$
[6] The conjugate of a binomial $a + b$ is $a - b$. Ex. Conjugate of $x + \sqrt{5}$ is $x - \sqrt{5}$.

we look at the radical expression and change the sign between the radical and the rest of the expression. Here, the conjugate of $\sqrt{x} - 2$ is $\sqrt{x} + 2$.

So now, $\lim_{x \to 4} \frac{\sqrt{x}-2}{x-4} = \lim_{x \to 4} \frac{\sqrt{x}-2}{x-4} \cdot \frac{\sqrt{x}+2}{\sqrt{x}+2} = \lim_{x \to 4} \frac{x-4}{(x-4)(\sqrt{x}+2)} = \lim_{x \to 4} \frac{1}{\sqrt{x}+2} = \frac{1}{4}$. Therefore,

$\lim_{x \to 4} \frac{\sqrt{x}-2}{x-4} = \frac{1}{4}$.

Ex. 7 Find the limit $\lim_{x \to 3} \frac{x-3}{\sqrt{x+6}-3}$.

<u>Solution:</u> Here, the conjugate of $\sqrt{x+6} - 3$ is $\sqrt{x+6} + 3$. So now, $\lim_{x \to 3} \frac{x-3}{\sqrt{x+6}-3} = \lim_{x \to 3} \frac{x-3}{\sqrt{x+6}-3} \cdot$

$\frac{\sqrt{x+6}+3}{\sqrt{x+6}+3} = \lim_{x \to 3} \frac{(x-3)(\sqrt{x+6}+3)}{x+6-9} = \lim_{x \to 3} \frac{(x-3)(\sqrt{x+6}+3)}{x-3} = \lim_{x \to 3} \sqrt{x+6} + 3 = \sqrt{3+6} + 3 = 6$.

Therefore, $\lim_{x \to 3} \frac{x-3}{\sqrt{x+6}-3} = 6$.

Simplifying Complex Fractions to Find Limits

Ex. 8 Find the limit $\lim_{x \to 0} \frac{x}{\frac{1}{x+2} - \frac{1}{2}}$.

<u>Solution:</u> Direct substitution doesn't work here, but we have a complex fraction which we should try to simplify first. We simplify complex fractions by multiplying the numerator and denominator by the least common denominator. The denominators of the "inside" fractions are $x + 2$ and 2, so the LCD is $2(x+2)$. Here we have $\lim_{x \to 0} \frac{x}{\frac{1}{x+2} - \frac{1}{2}} = \lim_{x \to 0} \frac{x}{\frac{1}{x+2} - \frac{1}{2}} \cdot \frac{2(x+2)}{2(x+2)} =$

$\lim_{x \to 0} \frac{2x(x+2)}{2-(x+2)} = \lim_{x \to 0} \frac{2x(x+2)}{-x} = \lim_{x \to 0} -2(x + 2) = -2(0 + 2) = -4$. Therefore, $\lim_{x \to 0} \frac{x}{\frac{1}{x+2} - \frac{1}{2}} = -4$.

Ex. 9 Find the limit $\lim\limits_{x \to 0} \dfrac{\frac{1}{x-1}+1}{x}$

Solution: Here there is just one inner denominator of x-1, so we will multiply the numerator and denominator of the outer fraction by x-1. $\lim\limits_{x \to 0} \dfrac{\frac{1}{x-1}+1}{x} = \lim\limits_{x \to 0} \dfrac{\frac{1}{x-1}+1}{x} \cdot \dfrac{x-1}{x-1} = \lim\limits_{x \to 0} \dfrac{1+x-1}{x(x-1)} =$

$\lim\limits_{x \to 0} \dfrac{x}{x(x-1)} = \lim\limits_{x \to 0} \dfrac{1}{x-1} = \dfrac{1}{0-1} = -1.$ Therefore, $\lim\limits_{x \to 0} \dfrac{\frac{1}{x-1}+1}{x} = -1.$

Using Trig Identities to Solve Limits

Ex. 10 Find the limit $\lim\limits_{\theta \to \frac{\pi}{2}} \dfrac{\sin^2(2\theta)}{1-\sin^2\theta}$

Solution: Direct substitution yields an indeterminate form of 0/0. We can use some trig identities to simplify the expression to find the limit. Notice that in the numerator we have a double angle, while in the denominator it appears that a Pythagorean identity may be useful. The double angle identity for *sin* is $\sin(2\theta) = 2\sin\theta\cos\theta$. The Pythagorean identity is $\sin^2\theta + \cos^2\theta = 1$. Let's continue: $\lim\limits_{\theta \to \frac{\pi}{2}} \dfrac{\sin^2(2\theta)}{1-\sin^2\theta} = \lim\limits_{\theta \to \frac{\pi}{2}} \dfrac{\sin^2(2\theta)}{1-\sin^2\theta} = \lim\limits_{\theta \to \frac{\pi}{2}} \dfrac{4\sin^2\theta\cos^2\theta}{1-\sin^2\theta} =$

$\lim\limits_{\theta \to \frac{\pi}{2}} \dfrac{4\sin^2\theta(1-\sin^2\theta)}{1-\sin^2\theta} = \lim\limits_{\theta \to \frac{\pi}{2}} 4\sin^2\theta = 4\sin^2\left(\dfrac{\pi}{2}\right) = 4.$ $\lim\limits_{\theta \to \frac{\pi}{2}} \dfrac{\sin^2(2\theta)}{1-\sin^2\theta} = 4.$

Ex. 11 Find the limit $\lim\limits_{\theta \to \frac{\pi}{2}} \dfrac{\cot^2\theta}{1-\sin\theta}$

Solution: Again, by direct substitution we end up with an indeterminate form 0/0. Here we have cotangent and sine. Usually it helps to rewrite all trig functions in terms of sine and cosine.

$$\lim_{\theta \to \frac{\pi}{2}} \frac{\cot^2 \theta}{1-\sin \theta} = \lim_{\theta \to \frac{\pi}{2}} \frac{\cos^2 \theta}{\sin^2 \theta (1-\sin \theta)} = \lim_{\theta \to \frac{\pi}{2}} \frac{\cos^2 \theta}{\sin^2 \theta (1-\sin \theta)} = \lim_{\theta \to \frac{\pi}{2}} \frac{1-\sin^2 \theta}{\sin^2 \theta (1-\sin \theta)} = \lim_{\theta \to \frac{\pi}{2}} \frac{(1+\sin \theta)(1-\sin \theta)}{\sin^2 \theta (1-\sin \theta)} =$$

$$\lim_{\theta \to \frac{\pi}{2}} \frac{(1+\sin \theta)}{\sin^2 \theta} = \frac{(1+\sin \frac{\pi}{2})}{\sin^2 \frac{\pi}{2}} = \frac{1+1}{1} = 2$$

Practice 3

1) $\lim\limits_{x \to 3} \dfrac{x^2 - 9}{x - 3}$

2) $\lim\limits_{x \to 2} \dfrac{x^2 - x - 2}{x - 2}$

3) $\lim\limits_{x \to -3} \dfrac{x + 3}{x^2 + 4x + 3}$

4) $\lim\limits_{x \to -1} \dfrac{x + 1}{x^2 - 2x - 3}$

5) $\lim\limits_{x \to 9} \dfrac{x - 9}{\sqrt{x} - 3}$

6) $\lim\limits_{x \to 16} \dfrac{\sqrt{x} - 4}{x - 16}$

7) $\lim\limits_{x \to 7} \dfrac{\sqrt{x - 6} - 1}{x - 7}$

8) $\lim\limits_{x \to 5} \dfrac{x - 5}{\sqrt{x + 4} - 3}$

9) $\lim\limits_{x \to 0} \dfrac{x}{\dfrac{1}{-2+x} + \dfrac{1}{2}}$

10) $\lim\limits_{x \to 0} \dfrac{\dfrac{1}{-1+x} + 1}{x}$

11) $\lim\limits_{x \to 0} \dfrac{\dfrac{1}{6+x} - \dfrac{1}{6}}{x}$

12) $\lim\limits_{x \to 0} \dfrac{\dfrac{1}{-5+x} + \dfrac{1}{5}}{x}$

13) $\lim\limits_{\theta \to \frac{\pi}{2}} \dfrac{1 - \sin^2 \theta}{\cos \theta}$

14) $\lim\limits_{\theta \to \pi} \dfrac{\cos^2 \theta - 1}{\sin(2\theta)}$

15) $\lim\limits_{\theta \to -\frac{\pi}{4}} \dfrac{1 + \sqrt{2}\sin \theta}{\cos(2\theta)}$

16) $\lim\limits_{\theta \to -\frac{\pi}{4}} \dfrac{\cos(2\theta)}{\cos \theta - \sin \theta}$

17) $\lim\limits_{\theta \to \frac{\pi}{2}} \dfrac{\sin(2\theta)}{\cos \theta}$

18) $\lim\limits_{\theta \to 0} \dfrac{\sin \theta}{\sin(2\theta)}$

19) $\lim\limits_{x \to -3} f(x),\ f(x) = \begin{cases} x^2 + 2x + 1, & x \neq -3 \\ 3, & x = -3 \end{cases}$

20) $\lim\limits_{x \to 1} f(x),\ f(x) = \begin{cases} 2x + 1, & x \neq 1 \\ 1, & x = 1 \end{cases}$

Squeeze Theorem (aka Sandwich Theorem)

Squeeze Theorem: If $f(x) \leq g(x) \leq h(x)$ for all x in some interval (c, d), except possibly at the point $a \in (c, d)$ and that $\lim_{x \to a} f(x) = \lim_{x \to a} h(x) = L$, for some number L.

Then it follows that $\lim_{x \to a} g(x) = L$.

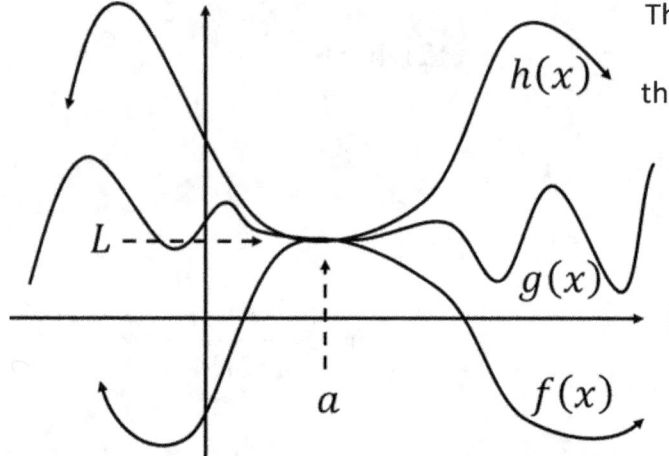

This picture demonstrates the squeeze theorem where g(x) is stuck between f(x) and h(x), so if we know the limits of f(x) and h(x) at a particular value a, then g(x) must also have the same limit at point a.

The squeeze theorem is somewhat difficult to apply in general because you need to find the right functions, but you should recognize its use and there are a couple of important limits that it proves.

Theorem[7]: $\lim_{x \to 0} \frac{\sin x}{x} = 1$

Theorem: $\lim_{x \to 0} \frac{1 - \cos x}{x} = 0$

Here's a couple of sample problems that tests your application of the squeeze theorem.

---

[7] Video proof available at https://www.khanacademy.org/math/ap-calculus-ab/ab-limits-new/ab-1-8/v/sinx-over-x-as-x-approaches-0 (Khan Academy, 2020)

Ex. 1 The function g is given by $g(x) = -2x^2 + 12x - 13$. The function h is given by h is given by $h(x) = x^2 - 6x + 14$. If f is a function that satisfies $g(x) \leq f(x) \leq h(x)$ for $0 < x < 5$, what is the $\lim_{x \to 3} f(x)$?

Solution: So, in this example we have an unknown function f(x) but we know some very important information. It is always between g(x) and h(x). So, if we want to find the limit $\lim_{x \to 3} f(x)$ we should check the limits $\lim_{x \to 3} g(x)$ and $\lim_{x \to 3} h(x)$ which both happen to be equal to 5. Therefore, $\lim_{x \to 3} f(x) = 5$ by the squeeze theorem.

Ex. 2 The function g is given by $g(x) = -x^2 + 4x - 2$. The function h is given by h is given by $h(x) = x^2 - 6x + 11$. If f is a function that satisfies $g(x) \leq f(x) \leq h(x)$ for $0 < x < 5$, what is the $\lim_{x \to 2} f(x)$?

Solution: This example looks very similar to the first, but when we check the limits, $\lim_{x \to 2} g(x) = 2$ and $\lim_{x \to 2} h(x) = 3$. Even though the function f is always between g and h, the condition that the limits are the same is not meant. Therefore, the sandwich theorem does not apply and $\lim_{x \to 2} f(x)$ cannot be determined.

In the next two examples we will use the fact that $\lim_{x \to 0} \frac{\sin x}{x} = 1$ or $\lim_{x \to 0} \frac{1-\cos x}{x} = 0$, and use basic trig identities and algebra to find the limits.

Ex. 3 Find the limit $\lim\limits_{x \to 0} \dfrac{\tan x}{2x}$

Solution: Our first step should be to rewrite everything in terms of sine and cosine. $\lim\limits_{x \to 0} \dfrac{\tan x}{2x} =$ $\lim\limits_{x \to 0} \dfrac{\sin x}{2x \cos x} = \lim\limits_{x \to 0} \dfrac{\sin x}{x} \cdot \dfrac{1}{2 \cos x} = \lim\limits_{x \to 0} \dfrac{\sin x}{x} \cdot \lim\limits_{x \to 0} \dfrac{1}{2 \cos x} = 1 \cdot \dfrac{1}{2} = \dfrac{1}{2}$. Therefore, $\lim\limits_{x \to 0} \dfrac{\tan x}{2x} = \dfrac{1}{2}$.

Ex. 4 Find the limit $\lim\limits_{x \to 0} \dfrac{\sin^2(3x)}{2x^2}$

Solution: Notice that these special limits work no matter what we substitute for x as long as the input of sine is the same as the denominator, for example: $\lim\limits_{x \to 0} \dfrac{\sin(7x)}{7x} = 1$ This one has a *3x* inside the sine, so we will need to adjust the denominator to make it work. $\lim\limits_{x \to 0} \dfrac{\sin^2(3x)}{2x^2} =$ $\lim\limits_{x \to 0} \dfrac{\sin(3x)\sin(3x)}{2x^2} = \lim\limits_{x \to 0} \dfrac{\sin(3x)\sin(3x)}{2x^2} \cdot \dfrac{9}{9} = \lim\limits_{x \to 0} \dfrac{\sin(3x)\sin(3x)}{3x \cdot 3x} \cdot \dfrac{9}{2} = \dfrac{9}{2} \lim\limits_{x \to 0} \dfrac{\sin(3x)}{3x} \cdot \lim\limits_{x \to 0} \dfrac{\sin(3x)}{3x} = \dfrac{9}{2} \cdot$ $1 \cdot 1 = \dfrac{9}{2}$

Ex. 5 Find the limit $\lim\limits_{x \to 0} \dfrac{1-\cos^2(4x)}{4x}$

Solution: Here we would use Pythagorean identity to simplify. $\lim\limits_{x \to 0} \dfrac{1-\cos^2(4x)}{4x} =$ $\lim\limits_{x \to 0} \dfrac{(1-\cos(4x))(1+\cos(4x))}{4x} = \lim\limits_{x \to 0} \dfrac{(1-\cos(4x))}{4x} \cdot \lim\limits_{x \to 0}(1+\cos(4x)) = 0 \cdot 2 = 0$. Therefore, $\lim\limits_{x \to 0} \dfrac{1-\cos^2(4x)}{4x} = 0$.

Practice 4

1) The function g is given by $g(x) = -2x^2 + 12x - 13$. The function h is given by $h(x) = x^2 - 6x + 14$. If f is a function that satisfies $g(x) \le f(x) \le h(x)$ for $0 < x < 5$, what is $\lim\limits_{x \to 3} f(x)$?

2) The function g is given by $g(x) = -2x^2 + 16x - 25$. The function h is given by $h(x) = x^2 - 8x + 23$. If f is a function that satisfies $g(x) \le f(x) \le h(x)$ for $0 < x < 5$, what is $\lim\limits_{x \to 3} f(x)$?

3) The function f is defined for all x in the interval $0 < x < 10$. Which of the following statements, if true, implies that $\lim\limits_{x \to 6} f(x) = 14$

A) There exist functions g and h with $f(x) \le g(x) \le h(x)$ for $0 < x < 10$, and $\lim\limits_{x \to 6} g(x) = \lim\limits_{x \to 6} h(x) = 14$

B) There exist functions g and h with $g(x) \le f(x) \le h(x)$ for $0 < x < 10$, and $\lim\limits_{x \to 6} g(x) = 13$ and $\lim\limits_{x \to 6} h(x) = 15$

C) There exist functions g and h with $g(x) \le f(x) \le h(x)$ for $0 < x < 10$, and $\lim\limits_{x \to 5} g(x) = \lim\limits_{x \to 7} h(x) = 14$

D) There exist functions g and h with $g(x) \le f(x) \le h(x)$ for $0 < x < 10$, and $\lim\limits_{x \to 6} g(x) = \lim\limits_{x \to 6} h(x) = 14$

4) The function f is defined for all x in the interval $0 < x < 10$. Which of the following statements, if true, implies that $\lim\limits_{x \to 8} f(x) = 6$

A) There exist functions g and h with $g(x) \le f(x) \le h(x)$ for $0 < x < 10$, and $\lim\limits_{x \to 8} g(x) = \lim\limits_{x \to 8} h(x) = 6$

B) There exist functions g and h with $g(x) \le h(x) \le f(x)$ for $0 < x < 10$, and $\lim\limits_{x \to 8} g(x) = 7$, $\lim\limits_{x \to 8} h(x) = 5$

C) There exist functions g and h with $g(x) \le f(x) \le h(x)$ for $0 < x < 10$, and $\lim\limits_{x \to 6} g(x) = \lim\limits_{x \to 6} h(x) = 8$

D) There exist functions g and h with $g(x) \le f(x) \le h(x)$ for $0 < x < 10$, and $\lim\limits_{x \to 7} g(x) = \lim\limits_{x \to 9} h(x) = 6$

5) $\lim\limits_{x \to 0} \dfrac{1 - \cos x}{x}$

6) $\lim\limits_{x \to 0} \dfrac{\sin x}{x}$

7) $\lim\limits_{x \to 0} \dfrac{5x}{\tan(4x)}$

8) $\lim\limits_{x \to 0} \dfrac{\sin(x)}{3x}$

9) $\lim\limits_{x \to 0} \dfrac{1 - \cos(2x)}{\cos(5x) - 1}$

10) $\lim\limits_{x \to 0} \dfrac{\sin(2x)\sin(5x)}{x\sin(4x)}$

Continuity

Intuitively, continuous functions can be drawn without lifting your pencil.

Definition: A function $f(x)$ is said to be continuous at $x = a$ if $\lim_{x \to a} f(x) = f(a)$. A function is said to be continuous in the interval $[c, d]$ if it continuous at each point in the interval.

Therefore, for a function to be continuous at a point three conditions must be met: 1) It must have a limit at that point, 2) It must be defined at that point, 3) The limit and the actual value of the function should match.

Fact: If $f(x)$ is continuous at $x = a$, then $\lim_{x \to a} f(x) = f(a)$ and $\lim_{x \to a^-} f(x) = f(a)$ and $\lim_{x \to a^+} f(x) = f(a)$

Ex. 1 Is the following function continuous?

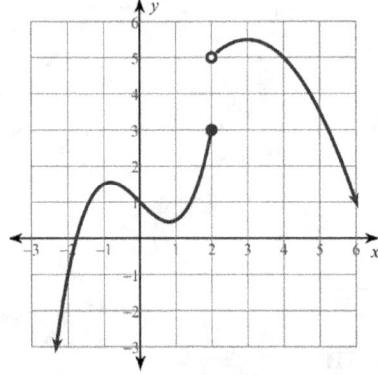

Solution: In this graph you can tell intuitively that it is not continuous because you would not be able to draw it without lifting your pencil. How do we prove it? Well the $\lim_{x \to 2^-} f(x) = 2$ and $\lim_{x \to 2^+} f(x) = 5$. The limit does not exist. Therefore, the function is not continuous at $x = 2$.

Ex. 2 Is the following function continuous?

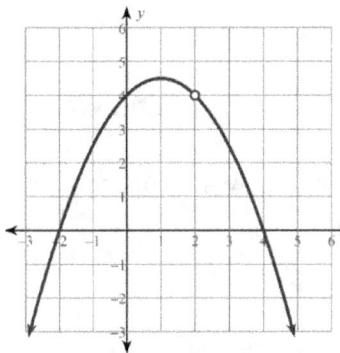

Solution: This graph almost looks like it can be drawn without lifting your pencil but notice that little open circle. That open circle represents a deleted or removed point. Let's prove that it really is discontinuous. Here the $\lim_{x \to 2^-} f(x) = 4$ and $\lim_{x \to 2^+} f(x) = 4$. So, the limit exists $\lim_{x \to 2} f(x) = 4$. But we still have a problem $f(2)$ is undefined. *Therefore, the function is not continuous at $x = 2$.*

Ex. 3 Where is the following function continuous?

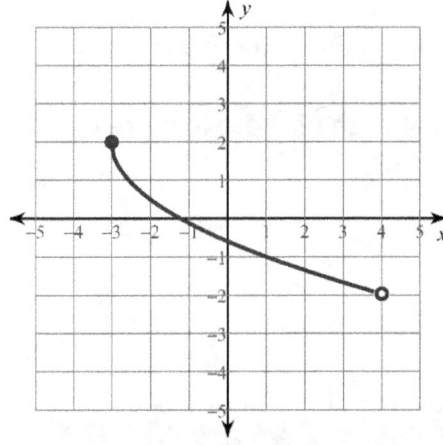

Solution: This function is continuous on the open interval (−3, 4). A function can only be continuous where it is defined and where the limit exists. Since the function is not defined at 4 and the limit is not defined at x = - 3, *the function is not continuous on the endpoints.*

Ex. 4 Is the following function continuous?

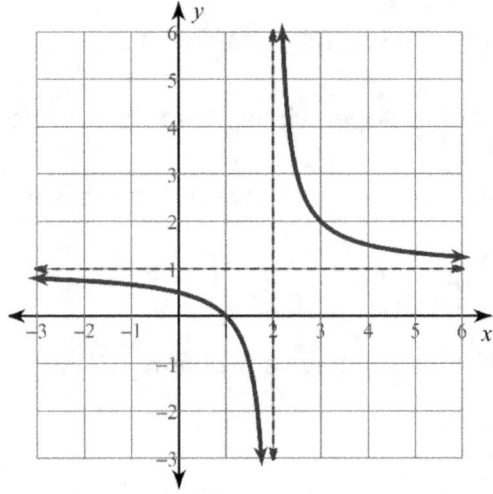

Solution: This graph is obviously discontinuous. It has the "worst" type of discontinuity, an asymptote. We can prove by taking limits on both sides as we approach x=2. Here the $\lim_{x \to 2^-} f(x) = -\infty$ and $\lim_{x \to 2^+} f(x) = \infty$. So, the limit exists $\lim_{x \to 2} f(x)$ does not exist. *Therefore, the function is not continuous at $x = 2$.*

Types of Discontinuity

There are four types of discontinuity: removable discontinuity, jump discontinuity, infinite discontinuity, and oscillating discontinuity. We will look at each of these in more detail:

Removable Discontinuity

Removable Discontinuity is basically when we delete a point from an otherwise continuous function and move it up or down. It is called a removable discontinuity because it can be removed or fixed easily by redefining the function to move the point back to the gap.

Definition: A function is said to have a removable discontinuity at a point *x=a* in its domain if $\lim_{x \to a} f(x) = L$ but $f(a)$ is equal to some other real number than L.

Fact: A removable discontinuity can be removed by redefining the function as $F(x) = \begin{cases} f(x), for\ x \neq a \\ L, for\ x = a \end{cases}$.

Here's a graph of a piecewise function $f(x) = \begin{cases} -\frac{1}{2}x + 2, x < 2 \\ 3, x = 2 \\ -3(x-3)^2 + 4, x > 2 \end{cases}$ to show what we are talking about.

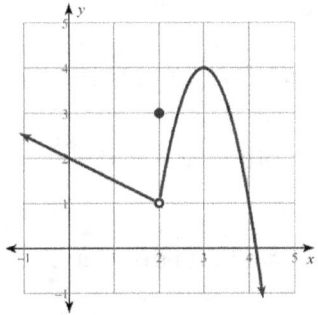

Removable Discontinuity

Jump Discontinuity

Jump Discontinuity is basically when the function jumps from one height to another height at a particular value of x = a. There's no easy fix to remove the discontinuity.

Definition: A function is said to have a jump discontinuity at a point x=a in its domain if $\lim_{x \to a^-} f(x) = L_1$ and $\lim_{x \to a^+} f(x) = L_2$, where L_1 and L_2 are real numbers and $L_1 \neq L_2$.

Here's a graph of a piecewise function $f(x) = \begin{cases} -\frac{1}{2}x + 1, x < 2 \\ -3(x-3)^2 + 4, x \geq 2 \end{cases}$

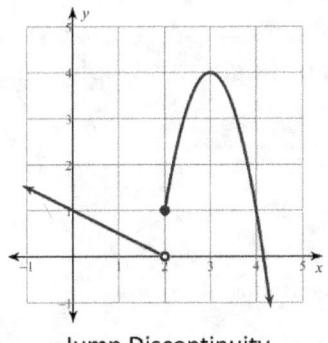

Jump Discontinuity

Infinite Discontinuity (Essential Discontinuity, Vertical Asymptote)

Infinite Discontinuity is basically when the function shoots up (or down) to infinity (or negative infinity) on at least one side of a particular value *x=a*. This also means there is a vertical asymptote at *x=a*. Functions that oscillate wildly are also sometimes classified as infinite discontinuity, but I prefer to classify them as their own type. These are also called essential discontinuities because they are the most extreme cases of discontinuity.

Definition: A function is said to have an infinite discontinuity at a point *x=a* in its domain if either $\lim_{x \to a^-} f(x)$ or $\lim_{x \to a^+} f(x)$ does not exist (or could be both).

Here's a graph of $f(x) = \frac{x-1}{x-2}$

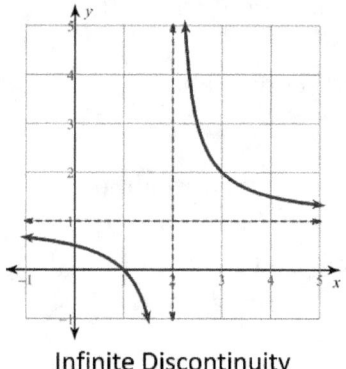

Infinite Discontinuity

Oscillating Discontinuity

Oscillating Discontinuity is basically when the function oscillates or fluctuates wildly when approaching a particular value *x = a*.

Here's a graph of $f(x) = \sin\frac{1}{x}$:

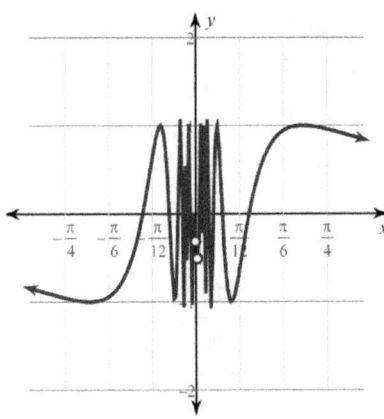

Oscillating Discontinuity

Summary of Discontinuities

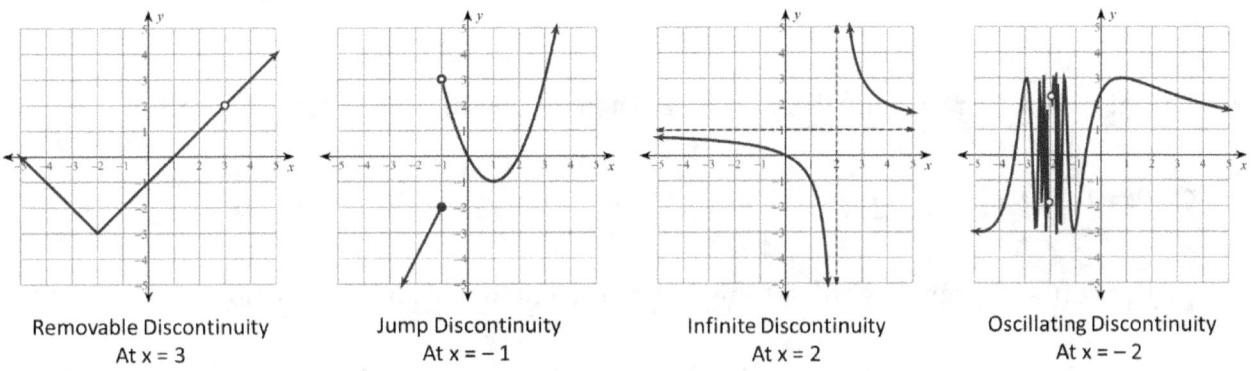

| Removable Discontinuity | Jump Discontinuity | Infinite Discontinuity | Oscillating Discontinuity |
| At x = 3 | At x = −1 | At x = 2 | At x = −2 |

Vertical Asymptotes and Limits

Definition: If $\lim_{x \to a^-} f(x) = \pm\infty$ or $\lim_{x \to a^+} f(x) = \pm\infty$, then we say that $f(x)$ has a vertical **asymptote at x=a.**

Usually when there is an asymptote, we see the function shoot up or down on both sides, but this definition implies there could be asymptotic behavior on just one side.

Here's an example of a function with a vertical asymptote on the right side of x = 2.

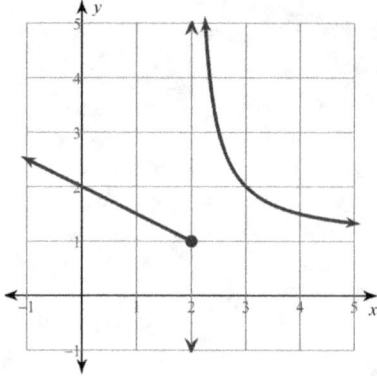

Horizontal Asymptotes and Limits

Definition: If $\lim_{x \to -\infty} f(x) = b$ or $\lim_{x \to \infty} f(x) = b$, then we say that $f(x)$ has a horizontal asymptote at y=b.

Horizontal asymptotes are determined by the end behavior of a function. The left and right end behaviors may be the same or they may be different. It is possible for a function to have two different horizontal asymptotes, one for end behavior on the left side and one for end behavior on the right side.

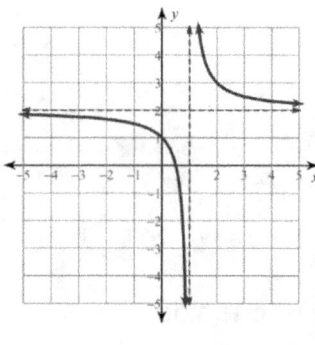

Horizontal asymptote
$y = 2$

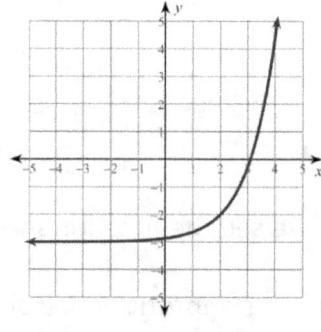

Horizontal asymptote
$y = -3$

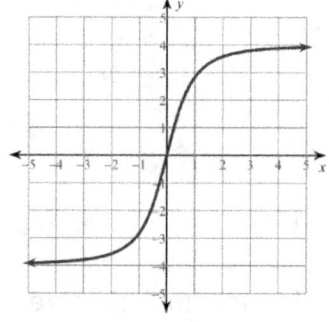

Horizontal asymptotes
$y = 4 \text{ and } y = -4$

Theorem: Polynomial[8] functions are continuous everywhere

Practice 5

1) Identify the discontinuity

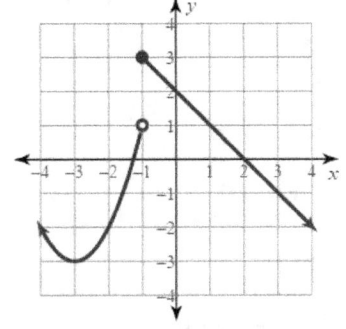

2) Identify the discontinuity

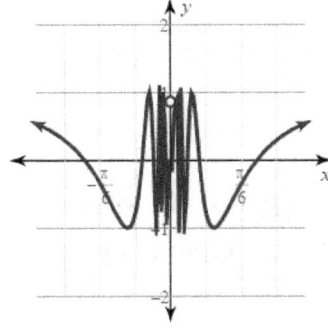

3) Identify the discontinuity

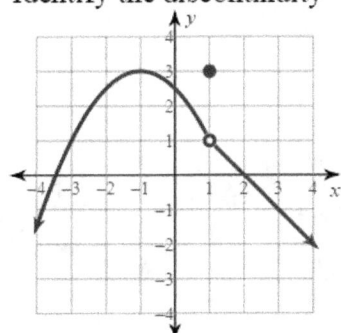

4) Identify the discontinuity

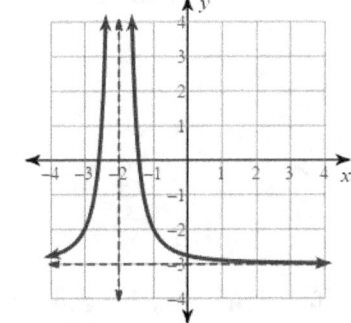

5) Find $\lim_{x \to 2} f(x)$

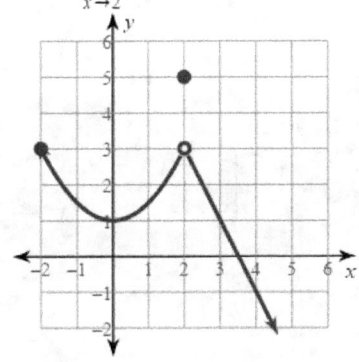

6) $f(x) = \begin{cases} \dfrac{|x|}{x}, & x \neq 0 \\ 0, & x = 0 \end{cases}$

Find $\lim_{x \to 0} f(x)$

---

[8]Polynomials $f(x) = a_n x^n + a_{n-1} x^{n-1} + \cdots + a_1 x^1 + a_0$ where a_i are constants and n_i are nonnegative integers.

7) $f(x) = \dfrac{x+3}{x^2+4x+3}$
On what intervals is the function $f(x)$ continuous?

8) $\begin{cases} 2^{x-1}, & x \leq 2 \\ -\dfrac{1}{2}x + b, & 2 < x \end{cases}$
Let f be the function defined above. For what values of b is f continuous at $x = 2$?

9) If $\lim\limits_{x \to 2} f(x) = \infty$, $\lim\limits_{x \to -\infty} f(x) = 6$, and $\lim\limits_{x \to 4} f(x) = 8$, then determine any asymptotes.

10) $g(x) = \begin{cases} \dfrac{x^2 - 2x - 8}{x - 4}, & x \neq 4 \\ k, & 4 = x \end{cases}$
Let g be the function defined above, where k is a constant. For what value of k is g continuous at $x = 4$?

11) $g(x) = \begin{cases} ax + a - x^2, & x < 4 \\ a + \dfrac{4}{x-3}, & x \geq 4 \end{cases}$
Let f be the function defined above. For what value of c, if any, is f continuous at $x = 4$?

12) The function g is defined by $g(x) = \dfrac{x^2 - 4}{x - 3}$
Find $\lim\limits_{x \to 3^-} g(x)$
Find $\lim\limits_{x \to 3^+} g(x)$

13) Evaluate the limit
$\lim\limits_{x \to -2} f(x)$, $f(x) = \begin{cases} x^2 + 6x + 6, & x \neq -2 \\ -4, & x = -2 \end{cases}$

14) Evaluate the limit
$\lim\limits_{x \to -2} f(x)$, $f(x) = \begin{cases} x^2 + 6x + 9, & x \leq -2 \\ -x - 1, & x > -2 \end{cases}$

15) $\lim\limits_{x \to -\infty} \dfrac{2x}{x+3}$

16) $\lim\limits_{x \to -\infty} \dfrac{16x}{x^2 + 16}$

17) $\lim\limits_{x \to -1} -\dfrac{x}{x+1}$

18) $\lim\limits_{x \to 3} -\dfrac{x-1}{x^2 - 4x + 3}$

19) $\lim\limits_{x \to -3^-} \dfrac{x+3}{x^2 + 4x + 3}$

20) $\lim\limits_{x \to -3} -\dfrac{3x}{x+3}$

21) Which of the following functions is continuous at $x = 1$?

  A) $f(x) = \begin{cases} x^2 + 2x - 5, & x < 1 \\ 2x - 4, & x \geq 1 \end{cases}$

  B) $f(x) = \dfrac{x+3}{x^2 + 2x - 3}$

  C) $f(x) = \begin{cases} -x^2 - 2x - 1, & x \leq 1 \\ -x - 1, & x > 1 \end{cases}$

  D) $f(x) = \begin{cases} 2x + 1, & x \leq 1 \\ -x^2 + 1, & x > 1 \end{cases}$

22) If $\lim\limits_{x \to 4} f(x)$ exists, with $\lim\limits_{x \to 4} f(x) < 6$ and $f(4) = 8$, which of the following statements must be false?

  A) $\lim\limits_{x \to 4^-} f(x) = 4$
  B) f is continuous at $x = 4$.
  C) $\lim\limits_{x \to 4^-} f(x) < 6$
  D) $\lim\limits_{x \to 4^-} f(x) = \lim\limits_{x \to 4^+} f(x)$

23) Let h be a function such that $\lim\limits_{x \to 4^-} h(x) = -\infty$. Which of the following statements must be true?

  A) h is undefined at 4
  B) There is a vertical asymptote at $x = -4$
  C) There is a vertical asymptote at $x = 4$
  D) $\lim\limits_{x \to 4^+} h(x) = -\infty$

24) If a function f is discontinuous which of the following must be false?

  A) $\lim\limits_{x \to 2^-} f(x) = \lim\limits_{x \to 2^+} f(x)$
  B) $\lim\limits_{x \to 2^-} f(x) = \lim\limits_{x \to 2} f(x)$
  C) $\lim\limits_{x \to 2} f(x) = f(2)$
  D) $\lim\limits_{x \to 2^-} f(x) = f(2)$

Intermediate Value Theorem

Intermediate Value Theorem (IVT): Suppose *f* is a function that is continuous on the closed interval [*a*, *b*]. If *L* is any number between *f*(*a*) and *f*(*b*), then there must be a value, *x* = *c*, where *a* < *c* < *b*, such that *f*(*c*) = *L*.

The word *intermediate* here means in the "middle" or "between". The intermediate value theorem basically states that if a function is continuous (unbroken) then when we go from a low point to a high point, we must pass through the points in the middle or between the low and high. For example, draw a straight line, then draw one point below the line and one point above the line. If you connect the points, then you must have crossed the line.

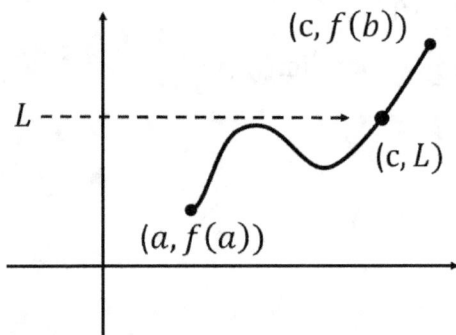

For the IVT to work, the function must be continuous. In the following graph, you can see why it may not work for a discontinuous function.

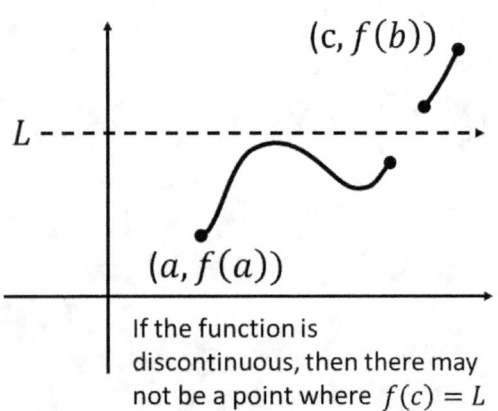

If the function is discontinuous, then there may not be a point where $f(c) = L$

One very special consequence of the IVT allows us to find zeros easily.

Intermediate Value Theorem (Zeros): Suppose *f* is a function that is continuous on the closed interval [*a*, *b*]. Suppose that *f*(*a*) and *f*(*b*) have opposite signs, then there must be a value, *x* = *c*, where *a* < *c* < *b*, such that *f*(*c*) = 0.

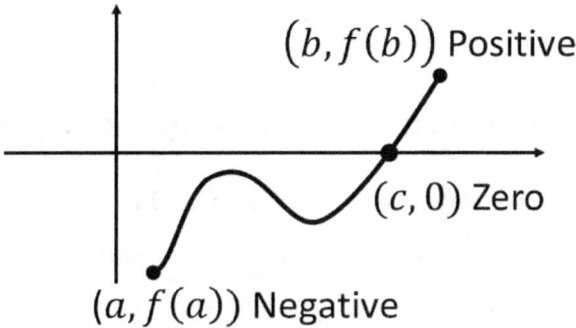

Ex. 1 Use the table below to answer the questions using IVT.

x	−6	−4	−2	0	2	4	6
$f(x)$	−6	4	2	10	−2	−8	6

A) There exists some x in the interval $[-6, -4]$ such that $f(x) = -4$

B) There exists some x in the interval $[-4, 0]$ such that $f(x) = -2$

C) There exists some x in the interval $[-2, 2]$ such that $f(x) = 0$

D) There exists at least one zero in the interval $[4, 6]$

E) There exists some x in the interval $[-2, 2]$ such that $f(x) = 8$

F) What is the minimum number of zeros between $[-6, 6]$

Solutions:

A) There exists some x in the interval $[-6,-4]$ such that $f(x) = -4$ **YES because -4 is between -6 and 4**

B) There exists some x in the interval $[-4, 0]$ such that $f(x) = -2$ **NO because -2 is not between 4 and 2 nor between 2 and 10**

C) There exists some x in the interval $[-2, 2]$ such that $f(x) = 0$ **YES because 0 is between 2 and -2**

D) There exists at least one zero in the interval $[4,6]$ **YES because 0 is between -8 and 6**

E) There exists some x in the interval $[-2, 2]$ such that $f(x) = 8$ **YES because 8 is between 2 and 10 and also between 10 and -2**

F) What is the minimum number of zeros between $[-6, 6]$ **3 at intervals (-6,-4), (0,2), and (4,6)**

Practice 6

Use the table below to answer the questions using IVT.

x	-6	-4	-2	0	2	4	6
$f(x)$	6	4	-2	8	-4	-8	6

1. There exists some x in the interval $[-6, -4]$ such that $f(x) = -4$

2. There exists some x in the interval $[-4, 0]$ such that $f(x) = 6$

3. There exists some x in the interval $[4,6]$ such that $f(x) = 0$

4. There exists at least one zero in the interval $[2,4]$

5. There exists some x in the interval $[-2, 2]$ such that $f(x) = -6$

6. What is the minimum number of zeros between $[-6, 6]$ and where do they occur?

Differentiation: Definition and Basic Rules

Differentiation is the process of finding the derivative which is one of the basic operations of calculus. The derivative gives information about the instantaneous rate of change at a particular point. Graphically, the derivative tells us the slope of a tangent line at that point. Calculus was discovered by a few different people around the same time. There are various ways to write the derivative including the following: $f'(x), y', \frac{dy}{dx}, D_x f, \dot{y}$. One of the most common being the Leibniz notation $\frac{dy}{dx}$ which emphasizes the derivative as a rate of change between a small change in y, dy with respect to a small change in x, dx. The other most used way being Lagrange notation $f'(x)$, which emphasizes the derivative being a function in its own right derived from the original function.

The derivative measures instantaneous rate of change. When we have a function that gives the position of an object, the derivative gives the rate of change of position, otherwise known as velocity. The derivative of velocity gives the acceleration.

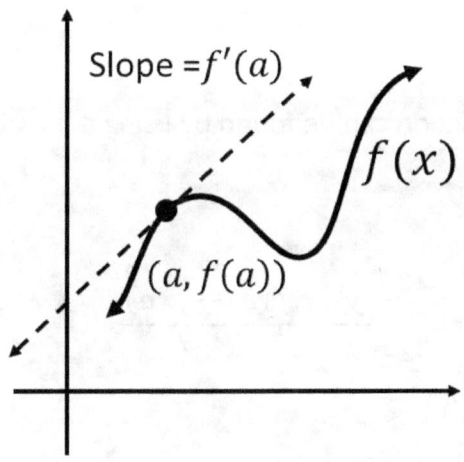

Average Rate of Change vs. Instantaneous Rate of Change

When you drive in your car and look at your speedometer, you are looking at your instantaneous velocity or instantaneous rate of change. However, many times while driving you may need to speed up, slow down or even come to a complete stop for some time. Instantaneous velocity is always changing. Your average velocity or average rate of change is your total distance divided by total time.

Visually this means that average rate of change will be the slope of a secant line through two points, while instantaneous rate of change will be the slope of a tangent line through one point.

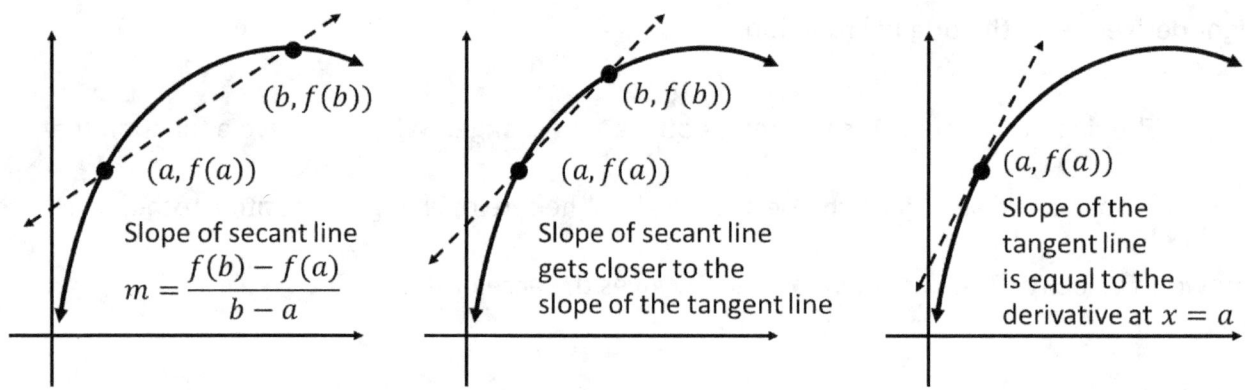

Average Rate of Change or the slope of a secant line of a function can be found by using a variation of the slope formula

$$Average\ Rate\ of\ Change = Slope\ of\ Secant\ Line = \frac{\Delta y}{\Delta x} = \frac{y_2 - y_1}{x_2 - x_1} = \frac{f(b) - f(a)}{b - a}$$

Definition: Average Rate of Change = $\frac{f(b)-f(a)}{b-a}$

Ex. 1 Find the average rate of change of $f(x) = x^2 + x + 1$ over the interval $[-3, -1]$.

Solution: So in this example b = − 1 and a = − 3. We need to find $f(-1) = (-1)^2 + (-1) + 1 = 1$ and $f(-3) = (-3)^2 + (-3) + 1 = 7$. Then we have $\frac{f(b)-f(a)}{b-a} = \frac{1-7}{-1--3} = \frac{-6}{2} = -3$.

Therefore, we have the average rate of change for the function is -3.

Ex. 2 What is the equation of the secant line that intersects the graph of $y = x^2 + 2x + 1$ with $x = -3$ and $x = -2$?

Solution: So in this example b = − 2 and a = − 3. We need to find $f(-2) = (-2)^2 + 2(-2) + 1 = 1$ and $f(-3) = (-3)^2 + (-3) + 1 = 4$. Then we have $\frac{f(b)-f(a)}{b-a} = \frac{1-4}{-2--3} = \frac{-3}{1} = -3$. The slope of the secant line will be -3. Now we use the point-slope form for the equation of a line using one of the points. $y - y_1 = m(x - x_1)$. We will use the first point (-3, 4) $y - 4 = -3(x - -3)$. then becomes $y - 4 = -3x - 9$ which if we put into slope-intercept form becomes $y = -3x - 5$. Therefore, the equation of the desired secant line is $y = -3x - 5$.

Estimating the Derivative at a Point

To estimate the derivative at a point we estimate the rate of change using information that we know that is closest to the point where we want the derivative.

Ex. 1 Estimate $f'(-10)$

x	−12	−9	−6	−4	−2	1	4
$f(x)$	2	−4	−1	−7	−3	6	9

Solution: Since we want to estimate the rate of change at x = − 10, the closest available x values are − 12 and − 9. This gives us two points to find the average rate of change closest to − 10. Plugging into our formula for average rate of change gives: $\frac{f(b)-f(a)}{b-a} = \frac{-4-2}{-9--12} = \frac{-6}{3} = -2$. Therefore, our estimate for $f'(-10) = -2$.

Ex. 2

Use the graph of $f(x)$ to estimate $f'(2)$

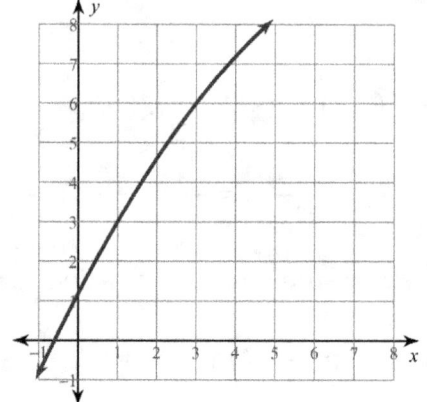

Solution: We want to estimate the derivative at x = 2 which is the same as estimating the slope of the tangent line at that point. We will use points close to the point, namely (1, 3) and (3, 6). Finding the slope or average rate of change between these two points gives: $\frac{f(b)-f(a)}{b-a} = \frac{6-3}{3-1} = \frac{3}{2}$. Therefore, our estimate for $f'(2) = \frac{3}{2}$.

Ex. 3

Estimate the equation of a tangent line to $f(x) = x^2 - x$ at the point $(2, 2)$

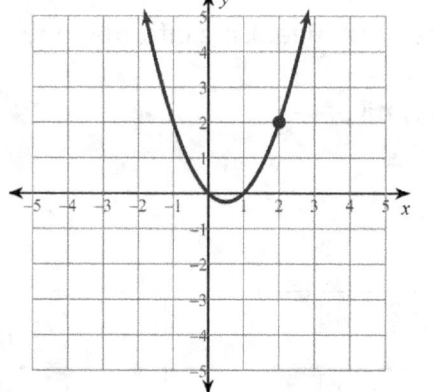

Solution: First, we want to estimate the derivative of $y = x^2 - x$ at (2,2). Notice that the point (1, 0) is also on the function. This means that the derivative or slope at x = 2 is steeper than 2. Perhaps a slope of 3 is a better estimate. If the slope is 3, then when we move down 6 and left 2, we would hit the y-axis at − 4. Therefore, our best estimate of the tangent line is $y = 3x - 4$.

Practice 7

1) Find the average rate of change of
   $f(x) = -2x^2 + 2$ over the interval $[-2, 1]$

2) Find the average rate of change of
   $f(x) = -\dfrac{1}{x+2}$ over the interval $[0, 3]$

3) What is the equation of the secant line that intersects the graph of
   $y = x^2 - 1$
   with $x = -2$ and $x = 0$

4) What is the equation of the secant line that intersects the graph of
   $y = \dfrac{1}{x+2}$
   with $x = 0$ and $x = 1$

5) Use the graph of $g(x)$ to compare $g'(1)$ and $g'(4)$

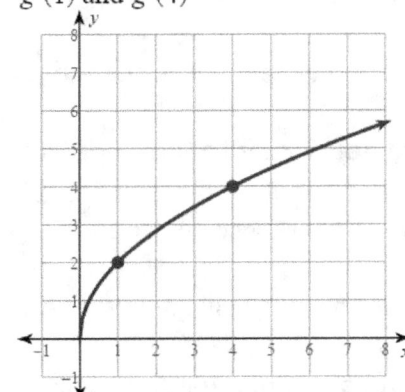

6) Use the graph of $h(x)$ to compare $h'(-3)$ and $h'(-2)$

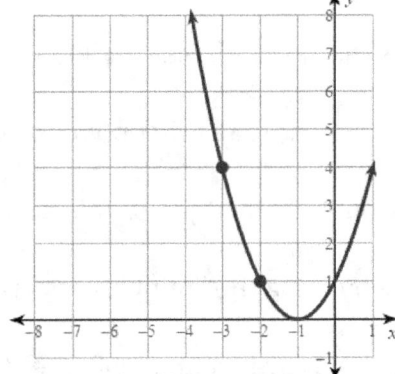

7) Use the graph of $f(x)$ to estimate $f'(3)$

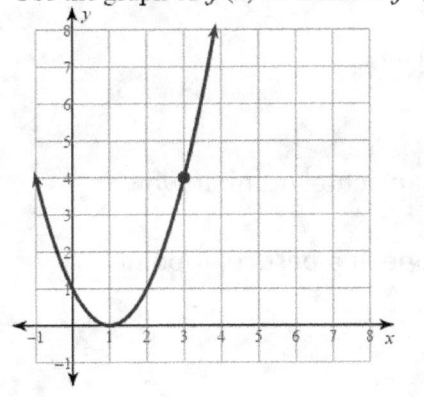

A) -4 B) 4
C) $\dfrac{1}{4}$ D) $-\dfrac{1}{4}$

8) Use the graph of $f(x)$ to estimate $f'(4)$

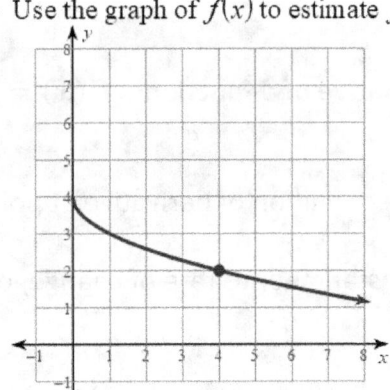

A) 4 B) -4
C) $\dfrac{1}{4}$ D) $-\dfrac{1}{4}$

9) Use the table below to estimate $f'(0)$

x	−12	−9	−6	−4	−2	1	4
f(x)	2	−4	−1	−7	−3	6	9

10) Use the table below to estimate $f'(-5)$

x	−12	−9	−6	−4	−2	1	4
f(x)	2	−4	−1	−7	−3	6	9

11) Use the table below to estimate $f'(-3)$

x	−12	−9	−6	−4	−2	1	4
f(x)	2	−4	−1	−7	−3	6	9

12) Use the table below to estimate $f'(3)$

x	−12	−9	−6	−4	−2	1	4
f(x)	2	−4	−1	−7	−3	6	9

Definition of the Derivative

The derivative is defined as the slope of a tangent line at one point which we approximate as the slope of a secant line through two points. When we allow the second point to get closer and closer to the first point, the slope of the secant gets closer to the slope of the tangent line through the first point. We approach the derivative. This is accomplished mathematically by using the limit process. Using the following definition[9], will give us a new function called the derivative that will give the instantaneous rate of change for any x-value that we wish to plug in.

Definition: Derivative of a function= $f'(x) = \lim_{h \to 0} \frac{f(x+h)-f(x)}{h}$

Using the definition of derivative at point, we will get an actual number which represents the instantaneous rate of change, or the actual slope at a particular point.

---

[9] Sometimes this definition is written as $f'(x) = \lim_{\Delta x \to 0} \frac{f(x+\Delta x)-f(x)}{\Delta x}$, where Δx "delta x" is a variable signifying a small change in x. I avoid this, because it is easy to confuse x with Δx.

Definition: Derivative at a point a = $f'(a) = \lim_{h \to 0} \frac{f(a+h)-f(a)}{h}$

The definition below gives an alternate way of setting up the problem, but still will give us the actual number.

Definition: (Alternate) Derivative at a point a = $f'(a) = \lim_{x \to a} \frac{f(x)-f(a)}{x-a}$

Using these limit definitions of the limit to find the derivative is quite tedious, but you should be able to use these definitions to find limits of some common limits, such as polynomials, rational functions, and square root functions. When using these definitions, there's often some "tricks" that we need to use to simplify these limits, the same tricks we used previously to evaluate limits.

Ex. 1 Find the derivative of $f(x) = x^2 + 3x + 4$. Then find $f'(2)$

Solution: When using the limit definitions to find the derivative of polynomials of quadratics and quadratics and higher degree, it's important to remember to distribute out correctly and FOIL.

Using the definition of the derivative, $f'(x) = \lim_{h \to 0} \frac{f(x+h)-f(x)}{h}$, we would have $f'(x) =$
$\lim_{h \to 0} \frac{(x+h)^2+3(x+h)+4-(x^2+3x+4)}{h} = \lim_{h \to 0} \frac{x^2+2xh+h^2+3x+3h+4-x^2-3x-4}{h}$ After simplifying the top this becomes $\lim_{h \to 0} \frac{2xh+h^2+3h}{h}$ after dividing by h becomes $\lim_{h \to 0} 2x + h + 3$ which when we take the limit becomes $2x + 3$. Therefore, $f'(x) = 2x + 3$. We can use this function to evaluate $f'(2) = 2(2) + 3 = 7$.

Using the definition of the derivative at a point, $f'(a) = \lim_{h \to 0} \frac{f(a+h)-f(a)}{h}$, we would have

$f'(2) = \lim_{h \to 0} \frac{(2+h)^2+3(2+h)+4-(x^2+3x+4)}{h} = \lim_{h \to 0} \frac{2^2+2(2)h+h^2+3(2)+3h+4-2^2-3(2)-4}{h}$. Again we

simplify the numerator to get $\lim_{h \to 0} \frac{4h+h^2+3h}{h}$ after dividing by h becomes $\lim_{h \to 0} 4+h+3$ which

when we take the limit becomes 7. *Therefore, we have the derivative at a specific point,*

$f'(2) = 7.$

Using the alternate definition of the derivative at a point, $f'(a) = \lim_{x \to a} \frac{f(x)-f(a)}{x-a}$, we

would have $f'(2) = \lim_{x \to 2} \frac{(x^2+3x+4)-(2^2+3(2)+4)}{x-2} = \lim_{x \to 2} \frac{x^2+3x+4-14}{x-2} = \lim_{x \to 2} \frac{x^2+3x-10}{x-2} =$

$\lim_{x \to 2} \frac{(x+5)(x-2)}{x-2} = \lim_{x \to 2} x + 5 = 7$. Therefore, using the alternate definition, we have, $f'(2) = 7.$

Ex. 2 Given $f(x) = \sqrt{x}$ Find the derivative of $f'(3)$

For limits with square roots, we will often have to use the conjugate of the radical expression to simplify the limit.

Using the definition of the derivative, $f'(x) = \lim_{h \to 0} \frac{f(x+h)-f(x)}{h}$, we would have $f'(x) =$

$\lim_{h \to 0} \frac{\sqrt{x+h}-\sqrt{x}}{h}$ we multiply the numerator and denominator by the conjugate of the radical

expression to get $\lim_{h \to 0} \frac{\sqrt{x+h}-\sqrt{x}}{h} \cdot \frac{\sqrt{x+h}+\sqrt{x}}{\sqrt{x+h}+\sqrt{x}} = \lim_{h \to 0} \frac{x+h-x}{h(\sqrt{x+h}+\sqrt{x})} = \lim_{h \to 0} \frac{h}{h(\sqrt{x+h}+\sqrt{x})} = \lim_{h \to 0} \frac{1}{\sqrt{x+h}+\sqrt{x}} =$

$\frac{1}{\sqrt{x+0}+\sqrt{x}} = \frac{1}{2\sqrt{x}}$. Therefore, $f'(x) = \frac{1}{2\sqrt{x}}$. We can use this function to evaluate $f'(3) = \frac{1}{2\sqrt{3}} =$

$\frac{\sqrt{3}}{6}$.

Using the definition of the derivative at a point, $f'(a) = \lim_{h \to 0} \frac{f(a+h)-f(a)}{h}$, we would have

$f'(3) = \lim_{h \to 0} \frac{\sqrt{3+h}-\sqrt{3}}{h} = \lim_{h \to 0} \frac{\sqrt{3+h}-\sqrt{3}}{h} \cdot \frac{\sqrt{3+h}+\sqrt{3}}{\sqrt{3+h}+\sqrt{3}} = \lim_{h \to 0} \frac{3+h-3}{h(\sqrt{3+h}+\sqrt{3})} = \lim_{h \to 0} \frac{h}{h(\sqrt{3+h}+\sqrt{3})} =$

$\lim_{h \to 0} \frac{1}{(\sqrt{3+h}+\sqrt{3})} = \frac{1}{\sqrt{3}+\sqrt{3}} = \frac{1}{2\sqrt{3}} = \frac{\sqrt{3}}{6}$. Therefore, here we have the derivative, $f'(3) = \frac{\sqrt{3}}{6}$.

Using the alternate definition of the derivative at a point, $f'(a) = \lim_{x \to a} \frac{f(x)-f(a)}{x-a}$, we

would have $f'(3) = \lim_{x \to 3} \frac{\sqrt{x}-\sqrt{3}}{x-3} = \lim_{x \to 3} \frac{\sqrt{x}-\sqrt{3}}{x-3} \cdot \frac{\sqrt{x}+\sqrt{3}}{\sqrt{x}+\sqrt{3}} = \lim_{x \to 3} \frac{x-3}{(x-3)(\sqrt{x}+\sqrt{3})} = \lim_{x \to 3} \frac{1}{\sqrt{x}+\sqrt{3}} = \frac{1}{\sqrt{3}+\sqrt{3}} =$

$\frac{1}{2\sqrt{3}} = \frac{\sqrt{3}}{6}$. Therefore, using the alternate definition, we have, $f'(3) = \frac{\sqrt{3}}{6}$.

Ex. 3 Given $f(x) = \frac{2}{x+5}$ Find the derivative of $f'(0)$.

Solution: When simplifying limits of rational functions, we will usually end up with a complex fraction that needs to be simplified by multiplying by the least common denominator.

Using the definition of the derivative, $f'(x) = \lim_{h \to 0} \frac{f(x+h)-f(x)}{h}$, we would have $f'(x) =$

$\lim_{h \to 0} \frac{\frac{2}{x+h+5} - \frac{2}{x+5}}{h}$ since we get a complex fraction we should multiply by the LCD of $(x + h +$

$5)(x + 5)$ in the numerator and denominator which gives $\lim_{h \to 0} \frac{\frac{2}{x+h+5} - \frac{2}{x+5}}{h} \cdot \frac{(x+h+5)(x+5)}{(x+h+5)(x+5)} =$

$\lim_{h \to 0} \frac{2(x+5)-2(x+h+5)}{h(x+5+h)(x+5)} = \lim_{h \to 0} \frac{2x+10-2x-2h-10}{h(x+5+h)(x+5)} = \lim_{h \to 0} \frac{-2h}{h(x+5+h)(x+5)} = \lim_{h \to 0} \frac{-2}{(x+5+h)(x+5)} =$

$\frac{-2}{(x+5)(x+5)} = \frac{-2}{(x+5)^2}$. Therefore, $f'(x) = \frac{-2}{(x+5)^2}$. We can use this function to evaluate $f'(0) =$

$\frac{-2}{(0+5)^2} = -\frac{2}{25}$.

Using the definition of the derivative at a point, $f'(a) = \lim_{h \to 0} \frac{f(a+h)-f(a)}{h}$, we would have

$$f'(0) = \lim_{h \to 0} \frac{\frac{2}{0+h+5} - \frac{2}{0+5}}{h} = \lim_{h \to 0} \frac{\frac{2}{h+5} - \frac{2}{5}}{h} \cdot \frac{5(h+5)}{5(h+5)} =$$

$$\lim_{h \to 0} \frac{2(5)-2(h+5)}{5h(h+5)} = \lim_{h \to 0} \frac{10-2h-10}{5h(h+5)} = \lim_{h \to 0} \frac{-2h}{5h(h+5)} = \lim_{h \to 0} \frac{-2}{5(h+5)} = -\frac{2}{25}.$$ Therefore, here we have the derivative, $f'(0) = -\frac{2}{25}$.

Using the alternate definition of the derivative at a point, $f'(a) = \lim_{x \to a} \frac{f(x)-f(a)}{x-a}$, we would have $f'(0) = \lim_{x \to 0} \frac{\frac{2}{x+5} - \frac{2}{5}}{x-0} = \lim_{x \to 0} \frac{\frac{2}{x+5} - \frac{2}{5}}{x} \cdot \frac{5(x+5)}{5(x+5)} = \lim_{x \to 0} \frac{10-2(x+5)}{5(x)(x+5)} =$

$\lim_{x \to 0} \frac{10-2x-10}{5(x)(x+5)} = \lim_{x \to 0} \frac{-2x}{5(x)(x+5)} = \lim_{x \to 0} \frac{-2}{5(x+5)} = -\frac{2}{25}$. Therefore, using the alternate definition, we have, $f'(0) = -\frac{2}{25}$.

Using the limit definition with trigonometric functions can be quite tricky. It requires using trig identities, usually the addition properties and the squeeze theorem (specifically the consequences that $\lim_{h \to 0} \frac{\sin h}{h} = 1$ and $\lim_{h \to 0} \frac{1-\cos h}{h} = 0$. Most of the time, you will not have to do trig derivatives using the definition, but here's one example:

<u>Ex. 4</u> Given $f(x) = \sin x$ Find the derivative of $f'(\pi)$

<u>Solution:</u> Using the definition of the derivative, $f'(x) = \lim_{h \to 0} \frac{f(x+h)-f(x)}{h}$, we would have

$f'(x) = \lim_{h \to 0} \frac{\sin(x+h) - \sin x}{h}$ We need to use addition identity for *sin* that gives

$\lim_{h \to 0} \frac{\cos x \sin h + \sin x \cos h - \sin x}{h} = \lim_{h \to 0} (\frac{\cos x \sin h}{h} + \frac{\sin x \cos h - \sin x}{h}) = \lim_{h \to 0} \frac{\cos x \sin h}{h} +$

$\lim_{h \to 0} \frac{\sin x (\cos h - 1)}{h} = \cos x \lim_{h \to 0} \frac{\sin h}{h} - \sin x \lim_{h \to 0} \frac{1 - \cos h}{h}$ We showed before using the squeeze

theorem that $\lim_{h \to 0} \frac{\sin h}{h} = 1$ and $\lim_{h \to 0} \frac{1 - \cos h}{h} = 0$. This gives us: $\cos x \cdot 1 - \sin x \cdot 0 = \cos x$.

Therefore, $f'(x) = \cos x$ and $f'(\pi) = \cos \pi = 1$.

Using the limit definition with logarithms can also be quite tricky. It requires using a substitution and properties of logs. Usually, we would not do this by hand but here's an example worked out.

Ex. 5 Given $g(x) = \ln x$ Find the derivative of $g'(e)$

Solution: This is one of the more difficult types of limits because it uses some tricky substitution as well as some properties of logs in order to solve. First, we will use the definition of the derivative, $g'(x) = \lim_{h \to 0} \frac{g(x+h) - g(x)}{h}$, we would have $g'(x) = \lim_{h \to 0} \frac{\ln(x+h) - \ln x}{h} =$

$\lim_{h \to 0} \frac{\ln \frac{x+h}{x}}{h} = \lim_{h \to 0} \frac{1}{h} \ln(1 + \frac{h}{x}) = \lim_{h \to 0} \ln\left(\left(1 + \frac{h}{x}\right)^{\frac{1}{h}} \right)$. We are going to make a substitution by letting

$n = \frac{h}{x}$. Then $h = nx$ which gives $\frac{1}{h} = \frac{1}{n} \cdot \frac{1}{x}$. As $h \to 0$ then $n \to 0$. This allows us to rewrite the

limit using these substitutions as $\lim_{n \to 0} \ln\left((1 + n)^{\frac{1}{n} \cdot \frac{1}{x}} \right)$ using the property of logs we can bring the

$\frac{1}{x}$ in front of the logarithm: $\lim_{n \to 0} \frac{1}{x} \ln\left((1 + n)^{\frac{1}{n}} \right)$. Since $\frac{1}{x}$ is not affected by the limit as n

approaches 0, we can take it completely out of the limit to get $\frac{1}{x} \lim_{n \to 0} \ln\left((1 + n)^{\frac{1}{n}} \right)$. Recall that

$\lim_{n \to 0} \left((1+n)^{\frac{1}{n}} \right) = e$, so taking the natural log of e gives 1. $\frac{1}{x} \lim_{n \to 0} \ln\left((1+n)^{\frac{1}{n}} \right) = \frac{1}{x}$. Therefore,

after all this algebraic manipulation we have $g'(x) = \frac{1}{x}$. Finally, $g'(e) = \frac{1}{e}$.

Practice 8

Use the definition of a derivative to find derivative $f'(x)$

1) $f(x) = 2x + 5$

2) $f(x) = 5x + 3$

3) $f(x) = 3x + 2$

4) $f(x) = -4x + 1$

5) $f(x) = 3x^2 + 2$

6) $f(x) = -2x^2 - 2$

7) $f(x) = 5x^2 - 3$

8) $f(x) = 4x^2 + 4$

9) $f(x) = 5x^2 + x + 2$

10) $f(x) = 3x^2 + 2x + 4$

11) $f(x) = 4x^2 + 4x + 2$

12) $f(x) = x^2 + 3x + 4$

13) $f(x) = \sqrt{x+5}$

14) $f(x) = \sqrt{2x+1}$

15) $f(x) = \sqrt{x-5}$

16) $f(x) = \sqrt{-3x+1}$

17) $f(x) = \dfrac{1}{x-4}$

18) $f(x) = -\dfrac{2}{x-1}$

19) $f(x) = \dfrac{2}{2x-3}$

20) $f(x) = -\dfrac{2}{2x+3}$

Differentiability

Differentiability is when we can take the derivative of a function at every interior point in a function's domain. Differentiability means that the graph will have a nonvertical tangent on every interior point. Differentiable functions can be described as "smooth," with no breaks, gaps, or sharp corners. Differentiable functions are also described as locally linear which means that if we zoom in (sometimes repeatedly) to a section, that section would appear like a line segment. We can see this on the calculator after repeating zooming in on the vertex of a quadratic function.

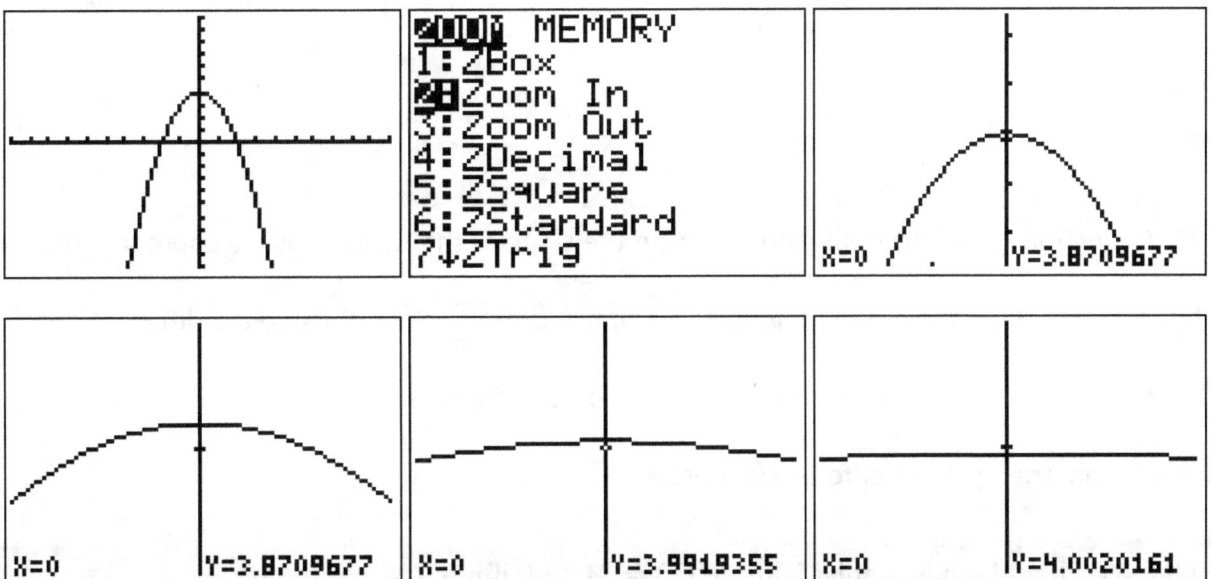

Definition: A function $f(x)$ is differentiable at $x = a$ if $\lim\limits_{h \to 0^-} \frac{f(a+h)-f(a)}{h} = \lim\limits_{h \to 0^+} \frac{f(a+h)-f(a)}{h}$

There are four different cases where a function could fail to be differentiable: If the function has a discontinuity, corner, cusp or vertical tangent it is not differentiable. Let's look at these more specifically:

Discontinuity

We already discussed continuity and the different types of discontinuity: jump discontinuity, removable discontinuity, infinite discontinuity, and oscillating discontinuity.

Examples of Discontinuity

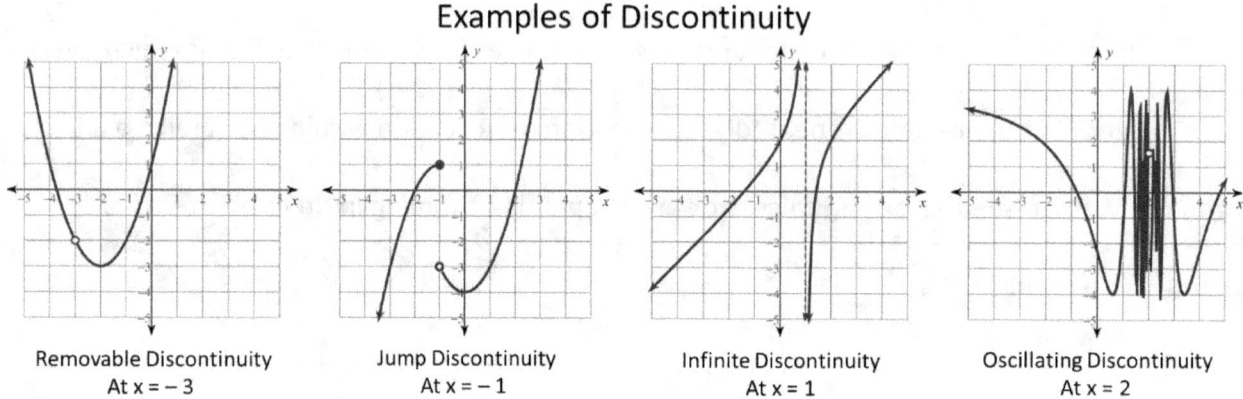

Removable Discontinuity
At x = −3

Jump Discontinuity
At x = −1

Infinite Discontinuity
At x = 1

Oscillating Discontinuity
At x = 2

Corner (Kink)

A function has a corner at a point x = a if the slope of the tangent as we approach from the left side and the slope of the tangent as we approach from the right side are different on each side. In other words, the derivative as we approach from the left and the derivative as we approach from the right are different values.

$f(x)$ has a corner at a if and only if $\lim_{x \to a^-} f'(x) = M$ and $\lim_{x \to a^+} f'(x) = N$.

Examples of Corners

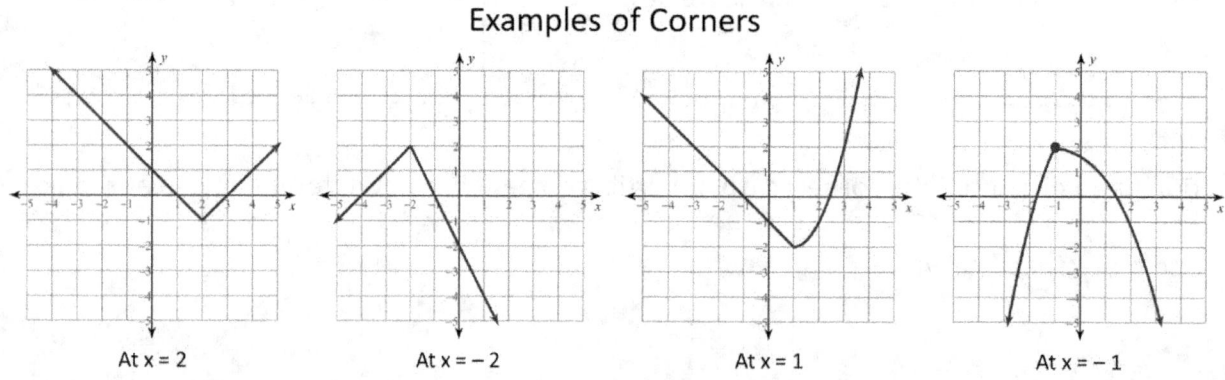

At x = 2

At x = −2

At x = 1

At x = −1

Vertical Tangent

A function has a vertical tangent at $x = a$ if the slope of the tangents as we approach $x = a$ approaches either infinity or negative infinity.

$f(x)$ has a vertical tangent at a if and only if $\lim_{x \to a} f'(x) = \infty$ or $\lim_{x \to a} f'(x) = -\infty$

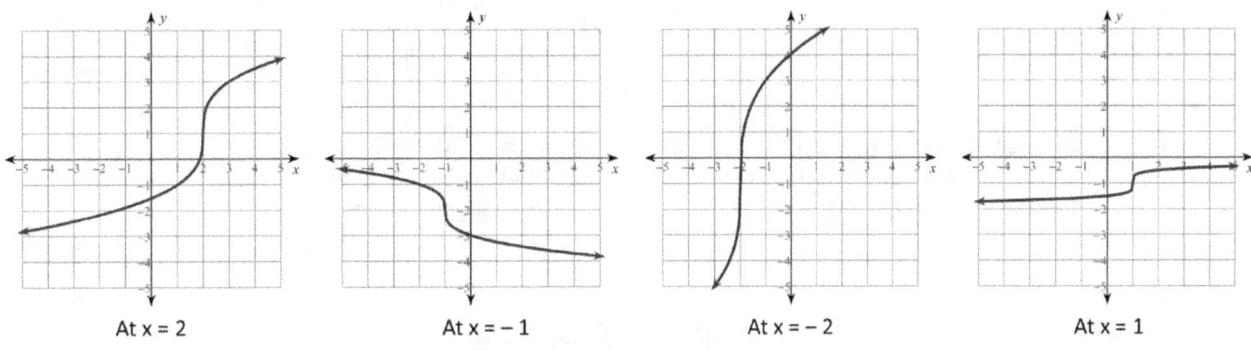

Examples of Vertical Tangents

At x = 2   At x = −1   At x = −2   At x = 1

Cusp

A function has a cusp at $x = a$ if the slope of the tangent as we approach from one side is infinity while the slope of the tangent as we approach from the other side is negative infinity.

$f(x)$ has a cusp at a if and only if:

1) $\lim_{x \to a^-} f'(x) = \infty$ while $\lim_{x \to a^+} f'(x) = -\infty$

   or

2) $\lim_{x \to a^-} f'(x) = -\infty$ while $\lim_{x \to a^+} f'(x) = \infty$

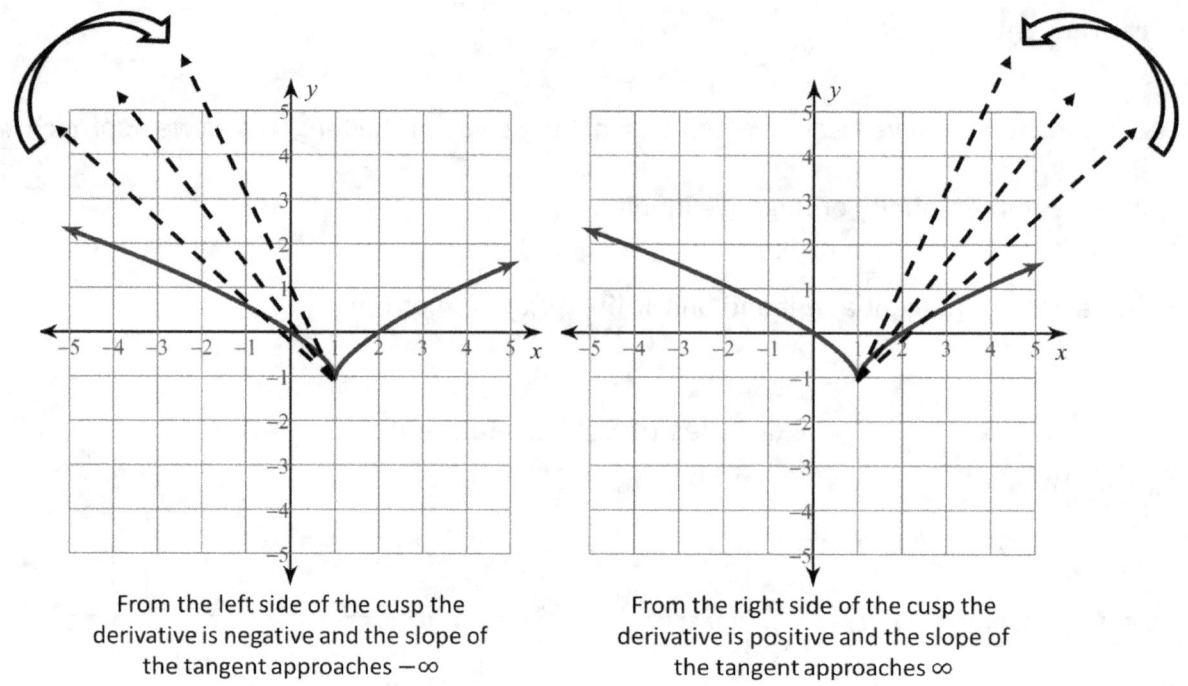

| From the left side of the cusp the derivative is negative and the slope of the tangent approaches $-\infty$ | From the right side of the cusp the derivative is positive and the slope of the tangent approaches ∞ |

Examples of Cusps

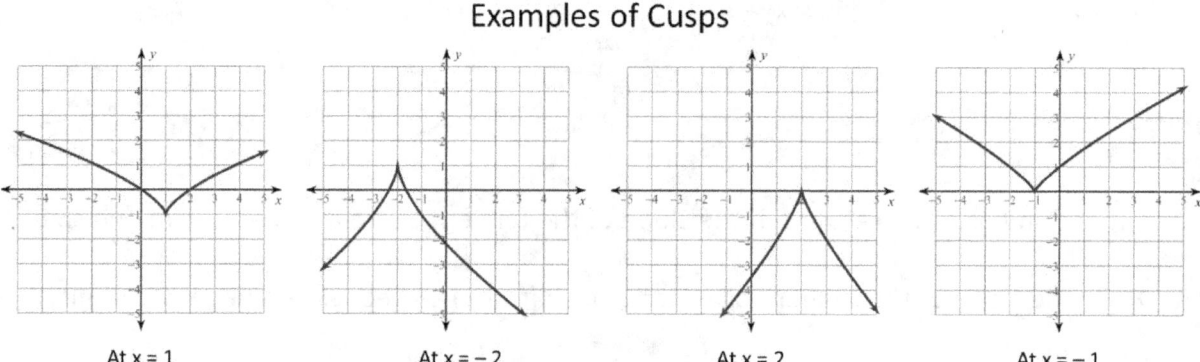

At x = 1   At x = − 2   At x = 2   At x = − 1

<u>Ex. 1</u> Where is the function differentiable?

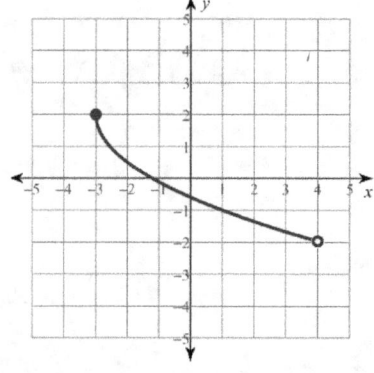

This function is defined on the interval [− 3, 4) but it differentiable only where the function is defined, and we can take the limit definition of the derivative. Therefore, the function is differentiable on the open interval (− 3, 4).

Ex. 2 Where is the function differentiable?

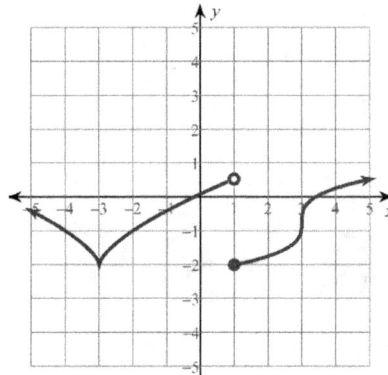

Solution: This function is defined for all reals. We have a cusp at $x = -3$, a jump discontinuity at $x = 1$, and a vertical tangent at $x = 3$. Therefore, this function is differentiable on the intervals: $(-\infty, -3) \cup (-3, 1) \cup (1, \infty)$.

Theorem: Polynomial[10] functions are differentiable everywhere

Practice 9

For each graph, determine if the function is differentiable. If not, classify why the derivative doesn't exist.

1)
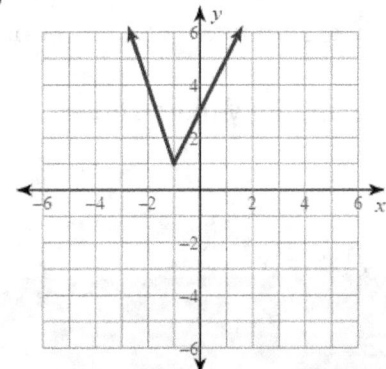

2)
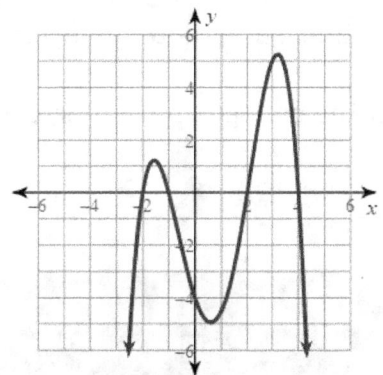

---

[10]Polynomials $f(x) = a_n x^n + a_{n-1} x^{n-1} + \cdots + a_1 x^1 + a_0$ where a_i are constants and n_i are nonnegative integers.

3)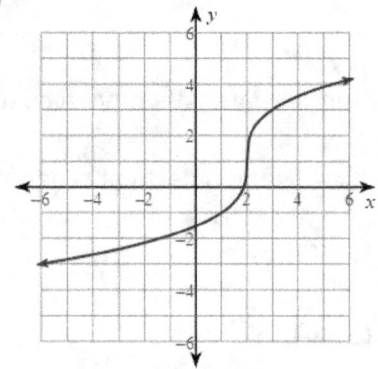

4)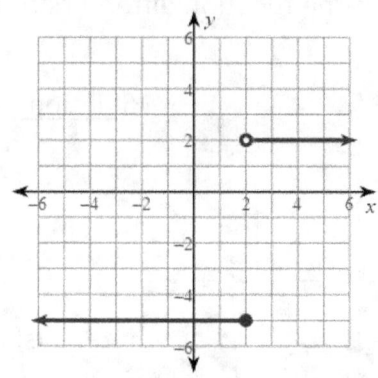

5)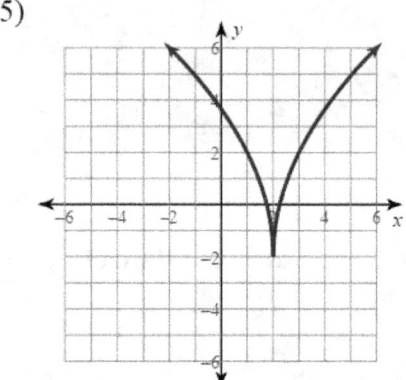

6)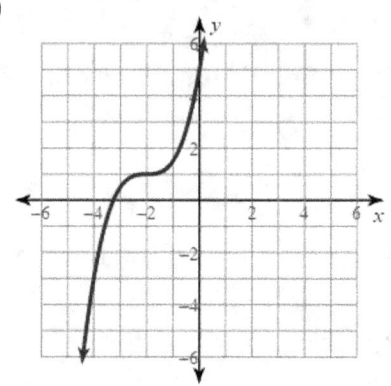

For each function below, determine if it is differentiable at *x = 0* by graphing with a calculator.

Tell whether the problem is a corner, cusp, vertical tangent, or discontinuity (name type).

7) $f(x) = \cot x$

8) $f(x) = x^{\frac{2}{3}}$

9) $f(x) = 2\sqrt[3]{x} - 2$

10) $f(x) = x^4 - 5x^2 + 4$

11) $f(x) = |x| - 4x$

12) $f(x) = \lfloor x \rfloor$

Determine if the following are true or false

13) If f is differentiable at $x = a$, the f is continuous at $x = a$.

14) If f is continuous at $x = a$, then f is differentiable at $x = a$.

15) If $\lim_{x \to a} f(x)$ exists then f is differentiable at $x = a$.

16) If f is differentiable at $x = a$, then the $\lim_{x \to a} f(x)$ exists.

17) If $\lim_{x \to a} f(x)$ exists, then f is continuous at a.

18) If f is continuous at a, then the $\lim_{x \to a} f(x)$ exists.

19) If f has a vertical tangent at a, then the f is not differentiable at $x = a$

20) If f has a horizontal tangent at a, then the f is not differentiable at $x = a$

21) If f is a polynomial function, then it is always differentiable.

22) If f is a absolute value function, then it is always differentiable.

Basic Derivative Rules

The limit definition can be used to find the derivative of any function, however it is very tedious to use, so we have some shortcut rules and formulas to find derivatives of the most common types of functions. Here's a list of common properties for derivatives:

Properties of Derivatives
1. Constant Rule: $f(x) = c$ then $f'(x) = 0$
2. Constant Multiple Rule: if $g(x) = c \cdot f(x)$ then $g'(x) = c \cdot f'(x)$
3. Power Rule: if $f(x) = x^n$ then $f'(x) = nx^{n-1}$
4. Sum or Difference: if $h(x) = f(x) \pm g(x)$ then $h'(x) = f'(x) \pm g'(x)$
5. Product Rule: if $h(x) = f(x)g(x)$ then $h'(x) = f(x)g'(x) + g(x)f'(x)$

6.	Quotient Rule: if $h(x) = \frac{f(x)}{g(x)}$ then $h'(x) = \frac{g(x)f'(x) - f(x)g'(x)}{g(x)^2}$
7.	Chain Rule: if $h(x) = f(g(x))$ then $h'(x) = f'(g(x))g'(x)$

We will go through how to use each of these properties.

Power Rule for Derivatives

The power rule along with the sum or difference allows us to take the derivative of any polynomial. The power rule tells us that to take the derivative, we take the "power" to the outside and drop it down by one.

$f(x)$		$f'(x)$
x^4	$\longrightarrow$	$4x^3$
x^3	$\longrightarrow$	$3x^2$
x^2	$\longrightarrow$	$2x$
$5x$	$\longrightarrow$	5
8	$\longrightarrow$	0

The power becomes the coefficient, then drop the power down by one.

<u>Ex. 1</u> What is the derivative of $f(x) = x^4 + 2x^3 - 4x^2 + 7x - 3$

<u>Solution:</u> The sum difference rule allows us to take the derivative term by term. The derivative of x^4 is $4x^3$. The derivative of $2x^3$ is $2 \cdot 3x^2$ or $6x^2$. The derivative of $-4x^2$ is $-4 \cdot 2x$ or $-8x$. The derivative of $7x$ is just 7. Lastly, the derivative of any constant is always 0. *This gives the final answer of* $f'(x) = 4x^3 + 6x^2 - 8x + 7$.

The power rule is one of the most important rules because it works for a lot of other type of functions including radicals and rational functions. Anything that can be written with exponents can be differentiated using the power rule.

Ex. 2 What is the derivative of $f(x) = \frac{2}{x^3} - \frac{5}{x^2} + \frac{3}{x} - 9$

Solution: Remember the property of negative exponents: $a^{-m} = \frac{1}{a^m}$. The first step in this type of problem is to rewrite using exponents $f(x) = 2x^{-3} - 5x^{-2} + 3x^{-1} - 9$. Again, we take the derivative of each term using the power rule. $f(x) = -6x^{-4} + 10x^{-3} - 3x^{-2}$. If needed, we can rewrite this as $f(x) = -\frac{6}{x^4} + \frac{10}{x^3} - \frac{3}{x^2}$.

Ex. 3 What is the derivative of $f(x) = \sqrt[4]{x} + 5\sqrt[3]{x^2} - 3\sqrt{x} + 5$

Solution: Recall the property of radicals and exponents: $\sqrt[n]{a^m} = a^{\frac{m}{n}}$. So, we can use this property to rewrite using exponents $f(x) = x^{\frac{1}{4}} + 5x^{\frac{2}{3}} - 3x^{\frac{1}{2}} + 5$. This allows us to take the derivative using the power rule. $f(x) = \frac{1}{4}x^{-\frac{3}{4}} + \frac{10}{3}x^{-\frac{1}{3}} - \frac{3}{2}x^{-\frac{1}{2}}$. We can rewrite back into radical form as $f(x) = \frac{1}{4\sqrt[4]{x^3}} + \frac{10}{3\sqrt[3]{x}} - \frac{3}{2\sqrt{x}}$.

Ex. 4 Find the equation of the line tangent to the function $f(x) = x^2 - 2x - 1$ at $x = -2$. Then find the equation of the normal line at the point $x = -2$.

Solution: To write the equation of a line, we need to know two things: a point and the slope. To find the point we simply evaluate the function at the given x value. $f(-2) = (-2)^2 - 2(-2) - 1 = 7$. This means that our point is $(-2, 7)$. Next we need to find the slope of the tangent. This is given by the derivative. Using the rules above we get $f'(x) = 2x - 2$. But we want the slope at x = – 2. We evaluate the derivative at $f'(-2) = 2(-2) - 2 = -6$. The slope of the tangent line is – 6. Recall the point-slope formula for a line to get: $y - 7 = -6(x - -2)$. Solving this for y and simplifying gives $y = -6x - 5$.

Normal lines are perpendicular to tangent lines. Slopes of perpendicular are opposite reciprocals $m_{norm} = -\frac{1}{m_{tan}}$. We still use the same point $(-2, 7)$. Again, use the point-slope formula for a line to get: $y - 7 = \frac{1}{6}(x - -2)$. Solving this for y and simplifying gives $y = \frac{x}{6} + \frac{22}{3}$.

Practice 10

Find the derivative of each

1) $f(x) = 5x^5 - x^3 + 1$

2) $f(x) = -3x^4 + 5x^2 - 3$

3) $y = 2x^4 + 4x^{-2} + 2x^{-4}$

4) $y = -x^5 + 5 + 2x^{-3}$

5) $f(x) = 1 + \dfrac{2}{x} + \dfrac{2}{x^4}$

6) $f(x) = 3 + \dfrac{1}{x^2} + \dfrac{4}{x^5}$

7) $f(x) = 5x^{\frac{5}{3}} + 2x^{\frac{3}{5}} + 4$

8) $y = 5x^{\frac{4}{5}} + 4x^{\frac{1}{5}} + 4$

9) $f(x) = \sqrt[3]{x} + \sqrt[5]{x} + 1$

10) $y = \sqrt[3]{x^2} + 2\sqrt[3]{x} - 2\sqrt[5]{x}$

11) $y = -2\sqrt[3]{x^2} - 3x^{\frac{1}{3}} + \dfrac{4}{x^4}$

12) $y = 4x^{\frac{1}{2}} + 5x^{-2} + 2x^{-3}$

13) $y = x^5 - 4x^{\frac{3}{2}} + 5\sqrt[3]{x^2}$

14) $y = -2x^{\frac{3}{5}} - 4x^{\frac{1}{2}} + 5x^{-2}$

15) $y = -4x^5 - 3x^{\frac{5}{3}} + \sqrt[3]{x^2}$

16) $y = 2x + x^{\frac{3}{4}} + \dfrac{3}{x^4}$

Product Rule for Derivatives

Product Rule: if $h(x) = f(x)g(x)$ then $h'(x) = f(x)g'(x) + g(x)f'(x)$

The product rule can be memorized as "first times derivative of the second plus second times derivative of the first" or FDS+SDF.

Ex. 1 Find the derivative of $f(x) = 5x^5(x^2 + 4)$

Solution: We could just multiply this out and then find the derivative, but first let's use the product rule. Our function is $f(x) = 5x^5(x^2 + 4)$. The first part is $5x^5$ and the second part is $(x^2 + 4)$. So using the product rule we have first $5x^5$ times derivative of the second $2x$, plus second $(x^2 + 4)$ times derivative of the first $25x^4$. When we put this together we have $5x^5(2x) + 25x^4(x^2 + 4)$. Remember for free response questions on the AP exam, you don't need to simplify your answer so you could just write: $5x^5(2x) + 25x^4(x^2 + 4)$. If we simplify this expression we get: $10x^6 + 25x^6 + 100x^4 = 35x^6 + 100x^4$.

Let's see what happens when we distribute first so we don't need to use this product rule. $f(x) = 5x^5(x^2 + 4) = 5x^7 + 20x^5$. Now we take the derivative to get $f'(x) = 35x^6 + 100x^4$. We get the same thing, so the product rule worked!

Practice 11

Find the derivative using the product rule

1) $f(x) = -2x^3(4x^3 + 1)$

2) $y = 3x^2(4x^5 - 4)$

3) $r = (2s^4 + 4)(-5s^2 + 2)$

4) $g(x) = (5x^2 + 3)(x^4 + 1)$

5) $h(w) = (-2w^2 - 3)(w^3 + 3w^2 - 2)$

6) $h(x) = (-2x^4 + 2x^3 + 5)(4x^4 + 1)$

7) $g = (-4t^4 - 2)(-3t^3 - 4t^2 + 1)$

8) $f = (3s^3 + 4s^2 + 4)(-3s^2 - 1)$

9) $f(x) = 3\sqrt[4]{x}(4x^4 - 5)$

10) $f(x) = \left(1 + \dfrac{3}{x^4}\right) \cdot -2x^3$

Quotient Rule for Derivatives

Quotient Rule: if $h(x) = \frac{f(x)}{g(x)}$ then $h'(x) = \frac{g(x)f'(x) - f(x)g'(x)}{g(x)^2}$

The quotient rule can be memorized as "bottom times derivative of the top minus top times derivative of the bottom all over bottom squared" or $\frac{BDT - TDB}{B^2}$.

<u>Ex. 1</u> Find the derivative of $f(x) = \frac{3x^2 + 4}{4x^4 + 5}$

Here the top (numerator) is $3x^2 + 4$ while the bottom (denominator) is $4x^4 + 5$. Using the quotient rule gives bottom $4x^4 + 5$ times derivative of the top $6x$ minus top $3x^2 + 4$ times derivative of the bottom $16x^3$. Put that together and make sure to put parentheses around each part, the numerator would be: $(4x^4 + 5)(6x) - (3x^2 + 4)(16x^3)$. The denominator becomes $(4x^4 + 5)^2$. This gives $\frac{(4x^4+5)(6x) - (3x^2+4)(16x^3)}{(4x^4+5)^2}$. Again, remember for free response questions on an AP exam, you don't need to simplify this mess. But if you need to simplify you would have to distribute and FOIL correctly to get: $\frac{(4x^4+5)(6x) - (3x^2+4)(16x^3)}{(4x^4+5)^2} =$

$\frac{24x^5 + 30x - 48x^5 - 64x^3}{16x^8 + 40x^4 + 25} = \frac{-24x^5 - 64x^3 + 30x}{16x^8 + 40x^4 + 25}$.

Practice 12

Find the derivative using the Quotient rule

1) $y = \dfrac{4}{x^4 + 5}$

2) $f(x) = \dfrac{4}{4x^2 - 3}$

3) $g(r) = \dfrac{2r^5 + 4}{4r^4 - 3}$

4) $s = \dfrac{5r^5 + 4r^4}{5r^3 + 4}$

5) $s = \dfrac{x^3 + 4x^2}{3x^2 + 4}$

6) $g(w) = \dfrac{w^3 + 5}{5w^3 + 4}$

7) $f(x) = \dfrac{2x^5 + x^3}{2x^4 - 5}$

8) $y = \dfrac{x^5 + 5}{4x^2 + 2}$

9) $y = \dfrac{3x^2 + 5}{3 + \dfrac{3}{x^2}}$

10) $y = \dfrac{4x^2 + 5}{2\sqrt[3]{x} - 4}$

Derivatives of Exponential and Logarithmic Functions

Here are the rules for the derivatives of exponential and logarithmic functions

$\dfrac{d}{dx} e^x = e^x$	$\dfrac{d}{dx} \ln x = \dfrac{1}{x}$
$\dfrac{d}{dx} a^x = a^x \ln a$	$\dfrac{d}{dx} \log_a x = \dfrac{1}{x \ln a}$

The exponential function is a unique function where the derivative is equal to itself.

Ex. 1 Use the limit definition to find the derivative of $f(x) = e^x$

<u>Solution:</u> Recall the limit definition of the derivative, $f'(x) = \lim\limits_{h \to 0} \dfrac{f(x+h) - f(x)}{h}$, we would have

$f'(x) = \lim\limits_{h \to 0} \dfrac{e^{x+h} - e^x}{h} = \lim\limits_{h \to 0} \dfrac{e^x e^h - e^x}{h} = \lim\limits_{h \to 0} \dfrac{e^x(e^h - 1)}{h}$ This allows us to take the e^x outside the

limit since it does not depend on h. $\lim\limits_{h \to 0} \dfrac{e^x(e^h - 1)}{h} = e^x \lim\limits_{h \to 0} \dfrac{e^h - 1}{h}$. We let $y = e^h - 1$, solving for

h gives $h = \ln(y + 1)$. This allows us to rewrite $e^x \lim\limits_{h \to 0} \dfrac{e^h - 1}{h}$ as $e^x \lim\limits_{y \to 0} \dfrac{y}{\ln(y+1)}$ which is the same

as $e^x \lim\limits_{y \to 0} \dfrac{1}{\frac{1}{y}\ln(y+1)}$ then using properties of logs gives $e^x \lim\limits_{y \to 0} \dfrac{1}{\ln(y+1)^{\frac{1}{y}}}$. Now we use the quotient rule and composition rule of limits allows us to rewrite this limit as $e^x \dfrac{1}{\ln \lim_{y \to 0}(y+1)^{\frac{1}{y}}}$ The

definition of $e = \lim_{n \to 0}(n+1)^{\frac{1}{n}}$ Finally we get, $e^x \dfrac{1}{\ln \lim_{y \to 0}(y+1)^{\frac{1}{y}}} = e^x \dfrac{1}{\ln e} = e^x$. Therefore, $\dfrac{d}{dx}e^x = e^x$.

We already presented how to use the limit definition to take the derivative of $\ln x$.

Ex. 2 Find the derivative of $f(x) = 3e^x + 2\ln x$

Solution: The derivative of e^x is e^x which is then multiplied by 3 to give $3e^x$. The derivative of $\ln x$ is $\dfrac{1}{x}$ and this is multiplied by 2 to give $\dfrac{2}{x}$. Add these together to get our final answer:

$f'(x) = 3e^x + \dfrac{2}{x}$.

Practice 13

1) $f(x) = 5e^x$
   Find $f'(x)$

2) $f(x) = -3e^x$
   Find $f'(x)$

3) $f(x) = 6\ln x$
   Find $f'(x)$

4) $f(x) = -5\ln x$
   Find $f'(x)$

5) $f(x) = 3^x$
   Find $f'(x)$

6) $f(x) = 10^x$
   Find $f'(x)$

7) $f(x) = 6\log_4 x$
   Find $f'(x)$

8) $f(x) = 3\log_7 x$
   Find $f'(x)$

9) $f(x) = \log x$
   Find $f'(x)$

10) $f(x) = 2\log x$
    Find $f'(x)$

Derivatives of Trigonometric Functions

$\frac{d}{dx}\sin x = \cos x$	$\frac{d}{dx}\cos x = -\sin x$
$\frac{d}{dx}\tan x = \sec^2 x$	$\frac{d}{dx}\cot x = -\csc^2 x$
$\frac{d}{dx}\sec x = \sec x \tan x$	$\frac{d}{dx}\csc x = -\csc x \cot x$

Notice in the table above, the similarities and differences in the formulas on the left side with those on the right side. Especially notice that all the *co-* functions (cosine, cotangent, cosecant) all have negative signs in their derivatives!

Ex. 1 Given $f(x) = \cos x$ Find the derivative of $f'(x)$

Solution: Using the definition of the derivative, $f'(x) = \lim_{h \to 0} \frac{f(x+h)-f(x)}{h}$, we would have

$f'(x) = \lim_{h \to 0} \frac{\cos(x+h)-\cos x}{h}$. We need to use addition identity for *cos* and the squeeze theorem[11]

giving us: $\lim_{h \to 0} \frac{\cos x \cos h - \sin x \sin h - \cos x}{h} = \lim_{h \to 0} \frac{\cos x \cos h - \cos x - \sin x \sin h}{h} = \lim_{h \to 0} (\cos x \frac{\cos h - 1}{h} -$

$\sin x \frac{\sin h}{h}) = \cos x \lim_{h \to 0} \frac{\cos h - 1}{h} - \sin x \lim_{h \to 0} \frac{\sin h}{h} = \cos x \,(0) - \sin x \,(1) = -\sin x$. Therefore,

$\frac{d}{dx} \cos x = -\sin x$.

Ex. 2 Given $g(x) = e^x \cos x$ Find the derivative of $g'(x)$.

Solution: Here we have the product of two functions which we know the derivative of each part. Using the product rule, we get first, e^x, times derivative of the second, $-\sin x$, plus second, $\cos x$, times derivative of the first, e^x. This gives: $g'(x) = -e^x \sin x + e^x \cos x$.

Ex. 3 Given $f(x) = \frac{\sin x}{1-\cos x}$ Find the derivative of $f'(x)$.

Solution: In this example, we must use the quotient rule to find the derivative. The top is $\sin x$ and the bottom is $1 - \cos x$. We take bottom, $1 - \cos x$, times derivative of the top, $\cos x$, minus top, $\sin x$, times the derivative of the bottom, $\sin x$. This gives us a numerator of $\cos x - \cos^2 x - \sin^2 x$. Of course this is all over bottom squared $(1 - \cos x)^2$. The answer becomes

---

[11] the squeeze theorem shows $\lim_{h \to 0} \frac{\sin h}{h} = 1$ and $\lim_{h \to 0} \frac{1-\cos h}{h} = 0$

$\frac{\cos x - \cos^2 x - \sin^2 x}{(1-\cos x)^2}$ If needed we could use a little trig and algebra to simplify further:

$$\frac{\cos x - \cos^2 x - \sin^2 x}{(1-\cos x)^2} = \frac{\cos x - (\cos^2 x + \sin^2 x)}{(1-\cos x)^2} = \frac{\cos x - 1}{(1-\cos x)^2} = \frac{\cos x - 1}{(\cos x - 1)^2} = \frac{1}{\cos x - 1}.$$

Once you know the derivatives for sine and cosine, you can solve for the derivatives of all the other trigonometric functions. You just need to express the other functions in term of sine and cosine, then use the quotient rule for derivatives.

<u>Ex. 4</u> Using the rules that $\frac{d}{dx}\sin x = \cos x$ and $\frac{d}{dx}\sin x = \cos x$, find the derivative for $f(x) = \tan x$

<u>Solution:</u> First, we must use the fact that $\tan x = \frac{\sin x}{\cos x}$. The top is $\sin x$ and the bottom is $\cos x$. We will use the quotient rule: bottom, $\cos x$, times derivative of the top, $\cos x$, minus top, $\sin x$, times derivative of bottom, $-\sin x$. This gives $\cos^2 x + \sin^2 x$, which goes over bottom squared. $\frac{\cos^2 x + \sin^2 x}{\cos^2 x}$. Recall the Pythagorean identity: $\sin^2 x + \cos^2 x = 1$. $\frac{\cos^2 x + \sin^2 x}{\cos^2 x} = \frac{1}{\cos^2 x} = \sec^2 x$. Therefore, $\frac{d}{dx}\tan x = \sec^2 x$

Practice 14

1) $f(x) = 4 + x^2 + \sin x$
   Find $f'(x)$

2) $f(x) = 3 - \frac{1}{x} + 2\cos x$
   Find $f'(x)$

3) $y = 6 - x^2 \cos x$

   Find $\dfrac{dy}{dx}$

4) $y = 4x + x \cot x$

   Find $\dfrac{dy}{dx}$

5) $y = x^3 \tan x$

   Find $\dfrac{dy}{dx}$

6) $y = x^4 \cot x$

   Find $\dfrac{dy}{dx}$

7) $y = \sqrt{x} \sec x$

   Find $\dfrac{dy}{dx}$

8) $y = \sqrt[3]{x} \csc x$

   Find $\dfrac{dy}{dx}$

9) $y = \dfrac{\cos x}{x^2}$

   Find $\dfrac{dy}{dx}$

10) $y = \dfrac{\sin x}{x^2}$

    Find $\dfrac{dy}{dx}$

11) $y = \dfrac{6}{\sin x}$

Find $\dfrac{dy}{dx}$

12) $f(x) = \dfrac{4}{\tan x}$

Find $f'(x)$

13) $f(x) = \dfrac{\sin x}{1 - \sin x}$

Find $f'(x)$

14) $f(x) = \dfrac{\cos x}{1 - \cos x}$

Find $f'(x)$

15) $f(x) = \sin^2 x$

Find $f'(x)$

16) $f(x) = \cos^2 x$

Find $f'(x)$

17) $y = -\tan(x)$

Find the slope at $x = -\dfrac{2\pi}{3}$

18) $y = -2\sin(x)$

Find the slope at $x = -\dfrac{2\pi}{3}$

19) $y = \cos(x)$

Find the slope at $x = -\dfrac{3\pi}{4}$

20) $y = -\csc(x)$

Find the slope at $x = \dfrac{\pi}{2}$

21) Find the equation of the line tangent to the function
$y = -2\sec(x)$ at $x = \pi$

22) Find the equation of the line tangent to the function
$y = -2\tan(x)$ at $x = 0$

23) Find the equation of the line tangent to the function
$y = \sin(x)$ at $x = -\dfrac{\pi}{3}$

24) Find the equation of the line tangent to the function
$y = -\sec(x)$ at $x = \dfrac{\pi}{6}$

Differentiation: Composite, Implicit, and Inverse

Chain Rule

Many functions are not just a single function but a composition of functions. Recall that a composition is when one function is plugged in or substituted into another function. We write the composition of functions as $f(g(x))$ or $f \circ g$ which is read "*f composed with g*" or "*f of g of x.*"

Chain Rule: if $h(x) = f(g(x))$ then $h'(x) = f'(g(x))g'(x)$

Basically, this states that when we have a composition of function with an "inside" function and an "outside" function, the derivative of the composition is the derivative of the outside (inside stays the same) times the derivative of the inside function.

<u>Ex. 1</u> Given that $h(x) = (3x^2 + 4x)^4$, find $h'(x)$

<u>Solution:</u> Here we have a composite function. The "inside" $f(x)$ is $3x^2 + 4x$ while the "outside" $g(x)$ function is x^4. To take the derivative we take the derivative of the outside keeping the inside the same like this: $4(3x^2 + 4x)^3$. Then we multiply this by the derivative of the inside: $6x + 4$. We can put this together to get: $4(3x^2 + 4x)^3(6x + 4)$. Usually we don't worry about simplifying the expression since expanding it out can be quite time consuming.

The chain rule can be used repeatedly if we have the composition of more than two functions. A function inside a function inside a function... We just keep chaining the derivatives together

Chain Rule for Three: if $m(x) = f(g(h(x)))$ then $m'(x) = f'(g(h(x)))g'(h(x))h'(x)$

Ex. 2 Given that $f(x) = ((2x^3 + 1)^2 + 3)^4$, find $f'(x)$

Solution: Here we have a function in a function in a function. Taking the derivative of the outside we get: $4((2x^3 + 1)^2 + 3)^3$. Multiply this by the derivative of the inside: $2(2x^3 + 1)$. Multiply this by the derivative of the inside of that: $6x^2$. Putting this altogether we get:

$4((2x^3 + 1)^2 + 3)^3 \cdot 2(2x^3 + 1) \cdot 6x^2$. We could simplify this a bit to get:

$48x^2((2x^3 + 1)^2 + 3)^3(2x^3 + 1)$

Ex. 3 Given the table below find the derivatives for each part

x	$f(x)$	$f'(x)$	$g(x)$	$g'(x)$
1	3	−1	4	−2
2	2	−1	2	$-\frac{3}{2}$
3	1	0	1	$\frac{1}{2}$
4	2	1	3	2

Part 1) Given $h_1(x) = f(x) + g(x)$, find $h_1'(2)$
Part 2) Given $h_2(x) = f(x) - g(x)$, find $h_2'(3)$
Part 3) Given $h_3(x) = f(x) \cdot g(x)$, find $h_3'(1)$
Part 4) Given $h_4(x) = \frac{f(x)}{g(x)}$, find $h_4'(4)$
Part 5) Given $h_5(x) = (f(x))^2$, find $h_5'(1)$
Part 6) Given $h_6(x) = f(g(x))$, find $h_6'(1)$

Solution: Part 1) $h_1(x) = f(x) + g(x)$. Using the sum rule for derivatives, gives us $h_1'(2) = f'(2) + g'(2)$. We look up the values for $f'(2)$ and $g'(2)$ in the table to get: $f'(2) + g'(2) = -1 + -\frac{3}{2} = -\frac{5}{2}$

Part 2) $h_2(x) = f(x) - g(x)$. Using the difference rule for derivatives, gives us $h_2'(3) = f'(3) - g'(3)$. We look up the values for $f'(3)$ and $g'(3)$ in the table to get: $f'(3) - g'(3) = 0 - \frac{1}{2} = -\frac{1}{2}$

Part 3) $h_3(x) = f(x) \cdot g(x)$. Using the product rule for derivatives, gives us $h_3'(1) = f(1) \cdot g'(1) + g(1) \cdot f'(1)$. We look up the values for $f(1)$, $f'(1)$, $g(1)$ and $g'(1)$ in the table to get: $f(1) \cdot g'(1) + g(1) \cdot f'(1) = 3 \cdot -2 + 4 \cdot -1 = -6 - 4 = -10$.

Part 4) $h_4(x) = \frac{f(x)}{g(x)}$. Using the quotient rule for derivatives, gives us $h_4'(4) = \frac{g(4) \cdot f'(4) - f(4) \cdot g'(4)}{(g(x))^2}$. We look up the values for $f(1)$, $f'(1)$, $g(1)$ and $g'(1)$ in the table to get:

$\frac{g(4) \cdot f'(4) - f(4) \cdot g'(4)}{(g(4))^2} = \frac{3 \cdot 1 - 2 \cdot 2}{3^2} = -\frac{1}{9}$.

Part 5) $h_5(x) = (f(x))^2$. Using the chain rule for derivatives, gives us $h_5'(1) = 2 \cdot f(1) \cdot f'(1)$. We look up the values for $f(1)$ and $f'(1)$ in the table to get: $2 \cdot f(1) \cdot f'(1) = 2 \cdot 3 \cdot -1 = -6$.

Part 6) $h_6(x) = f(g(x))$. Using the chain rule for derivatives, gives us $h_6'(1) = f'(g(1)) \cdot g'(1)$. We look up the values for $g(1)$ and $g'(1)$ in the table to get: $f'(g(1)) \cdot g'(1) = f'(4) \cdot g'(1)$. We then look up the value for $f'(4)$ to get: $f'(4) \cdot g'(1) = 1 \cdot -2 = -2$.

Practice 15

1) $y = (4x^3 + 1)^4$

2) $y = (2x^3 + 5)^3$

3) $y = \sqrt[3]{5x^3 - 4}$

4) $y = \sqrt[4]{x^5 + 5}$

5) $y = \dfrac{1}{(3x^2 + 1)^3}$

6) $y = \dfrac{1}{(5x^3 - 3)^2}$

7) $y = \left((5x^4 + 3)^5 - 3\right)^3$

8) $y = \left((3x^5 - 2)^2 - 5\right)^5$

9) $y = \sin 3x^2$

10) $y = \cos 3x^5$

11) $y = e^{2x^4}$

12) $y = e^{3x^4}$

13) $y = e^{x^2}$

14) $y = e^{3x^5}$

15) $y = (x^5 + 5)(3x^4 - 2)^5$

16) $y = (-2x^2 + 5)(2x + 1)^3$

17) $y = \dfrac{2x + 5}{(4x^4 + 5)^2}$

18) $y = \dfrac{(x^2 - 2)^2}{5x^3 + 4}$

19)

x	$f(x)$	$f'(x)$	$g(x)$	$g'(x)$
1	2	1	1	2
2	3	1	3	$\frac{3}{2}$
3	4	$-\frac{1}{2}$	4	0
4	2	-2	3	-1

Part 1) Given $h_1(x) = f(x) + g(x)$, find $h_1'(3)$
Part 2) Given $h_2(x) = f(x) - g(x)$, find $h_2'(3)$
Part 3) Given $h_3(x) = f(x) \cdot g(x)$, find $h_3'(2)$
Part 4) Given $h_4(x) = \frac{f(x)}{g(x)}$, find $h_4'(2)$
Part 5) Given $h_5(x) = (f(x))^2$, find $h_5'(1)$
Part 6) Given $h_6(x) = f(g(x))$, find $h_6'(3)$

20)

x	$f(x)$	$f'(x)$	$g(x)$	$g'(x)$
1	2	2	3	-2
2	4	$\frac{1}{2}$	1	0
3	3	-1	3	$\frac{3}{2}$
4	2	-1	4	1

Part 1) Given $h_1(x) = f(x) + g(x)$, find $h_1'(2)$
Part 2) Given $h_2(x) = f(x) - g(x)$, find $h_2'(1)$
Part 3) Given $h_3(x) = f(x) \cdot g(x)$, find $h_3'(4)$
Part 4) Given $h_4(x) = \frac{f(x)}{g(x)}$, find $h_4'(1)$
Part 5) Given $h_5(x) = (f(x))^2$, find $h_5'(1)$
Part 6) Given $h_6(x) = f(g(x))$, find $h_6'(1)$

Implicit Differentiation

Usually when we work with equations we have the equation defined explicitly or directly in the form of having y = on the left side of the equation and some expression with x on the right side, for example $y = x^2 + 3x - \cos x$. We can differentiate these equations explicitly by using all the formulas and rules we have gone over. But not all equations are written this way. In some equations we may not be able to solve for y directly. For example, the equation of a circle: $x^2 + y^2 = 9$. For these types of equations, we need to differentiate implicitly. When we differentiate implicitly, we consider y to be a function of x and since we are taking the derivative of y with respect to x, $\frac{dy}{dx}$, then every time we differentiate a term with y, we will need to use the chain rule.

<u>Ex. 1</u> Find the derivative $\frac{dy}{dx}$ given that $2x^3 + 4y^3 = 7$

<u>Solution:</u> We apply the derivative operator to both sides to get: $\frac{d}{dx}[2x^3 + 4y^3] = \frac{d}{dx}[7]$. On the left side we have a sum and on the right side we have a constant: $\frac{d}{dx}[2x^3] + \frac{d}{dx}[4y^3] = 0$. For the first term we use the power rule. For the second term we need to use the chain rule and think of y as a function of x: $6x^2 + 12y^2 \cdot \frac{dy}{dx} = 0$. Now we want to get the $\frac{dy}{dx}$ by itself: $12y^2 \cdot \frac{dy}{dx} = -6x^2$. Divide to get: $\frac{dy}{dx} = -\frac{6x^2}{12y^2}$. Simplifying gives: $\frac{dy}{dx} = -\frac{x^2}{2y^2}$.

<u>Ex. 2</u> Find the derivative $\frac{dy}{dx}$ given that $4x^3 + 3y = 5x^2y^2$

<u>Solution:</u> We apply the derivative operator to both sides to get: $\frac{d}{dx}[4x^3 + 3y] = \frac{d}{dx}[5x^2y^2]$. On the left side we have a sum and on the right side we have a product of $5x^2$ and y^2. Using the sum rule on the left side and product rule on the right: $\frac{d}{dx}[4x^3] + \frac{d}{dx}[3y] = 5x^2\frac{d}{dx}[y^2] + y^2\frac{d}{dx}[5x^2]$. Continuing: $12x^2 + 3\frac{dy}{dx} = 5x^2\frac{d}{dx}[y^2] + y^2\frac{d}{dx}[5x^2]$. This gives: $12x^2 + 3\frac{dy}{dx} =$

$5x^2 \cdot 2y\frac{dy}{dx} + y^2 \cdot 10x$. Simplifying a little: $12x^2 + 3\frac{dy}{dx} = 10x^2y\frac{dy}{dx} + 10xy^2$. We have $\frac{dy}{dx}$ in more than one spot. We need to get the $\frac{dy}{dx}$s to the left side and get the remaining terms to the right side. This gives: $3\frac{dy}{dx} - 10x^2y\frac{dy}{dx} = 10xy^2 - 12x^2$. We factor out the $\frac{dy}{dx}$ to give: $\frac{dy}{dx}(3 - 10x^2y) = 10xy^2 - 12x^2$. Dividing gives us the final answer: $\frac{dy}{dx} = \frac{10xy^2 - 12x^2}{3 - 10x^2y}$

Ex. 3 Find the equation of the tangent line of $x^3 + 5y^3 = 3$ at the point $(2, -1)$

Solution: To write the equation of a line we need to know a point and the slope. The slope is going to be the slope of the tangent which we can find by taking the derivative. We can find the derivative implicitly: $\frac{dy}{dx}x^3 + \frac{dy}{dx}5y^3 = \frac{dy}{dx}3 \Rightarrow 3x^2 + 15y^2\frac{dy}{dx} = 0$. Solving for the derivative: $\frac{dy}{dx} = -\frac{x^2}{5y^2}$. We can use this to find the slope of the tangent at $(2, -1)$. Plugging 2 in for x and -1 in for y: $\frac{dy}{dx} = -\frac{(2)^2}{5(-1)^2} = -\frac{4}{5}$. Using the point-slope formula gives us: $y - (-1) = -\frac{4}{5}(x - 2)$. Simplifying: $y = -\frac{4}{5}x + \frac{3}{5}$.

Ex. 4 Find the second derivative $\frac{d^2y}{dx^2}$ of $3x^2 = -2y^2 + 4$

Solution: Taking the first derivative implicitly: $\frac{dy}{dx}[3x^2] = \frac{dy}{dx}[-2y^2] + \frac{dy}{dx}[4] \Rightarrow 6x = -4y\frac{dy}{dx}$.

Then, the first derivative is: $\frac{dy}{dx} = -\frac{3x}{2y}$. Now we take the derivative of this to get the second derivative: $\frac{d^2y}{dx^2} = \frac{(2y)(-3)-(-3x)(2\frac{dy}{dx})}{(2y)^2} = \frac{-6y+6x\frac{dy}{dx}}{4y^2}$. Notice that this expression for the second derivative includes the first derivative, but we know that $\frac{dy}{dx} = -\frac{3x}{2y}$, so we substitute this in to

get: $\frac{-6y+6x(-\frac{3x}{2y})}{4y^2} = \frac{-6y-\frac{9x^2}{y}}{4y^2}$. Simplify this complex fraction: $\frac{-6y-\frac{9x^2}{y}}{4y^2} = \frac{-6y^2-9x^2}{4y^3}$.

Practice 16

1) $-y^2 + 2 = 5x^2$
   Find $\frac{dy}{dx}$

2) $3x^2 + 5y^3 = 1$
   Find $\frac{dy}{dx}$

3) $3x^3 + 4y = y^2$
   Find $\frac{dy}{dx}$

4) $4x^2 = -4y + 2y^2$
   Find $\frac{dy}{dx}$

5) $3x^3 = -4x^3y^3 + 5$
   Find $\frac{dy}{dx}$

6) $1 = 5x^3 - 5xy^2$
   Find $\frac{dy}{dx}$

7) $3x^3 + 3y^2 = 5xy^2$

Find $\dfrac{dy}{dx}$

8) $2y^2 = 3x + x^2y^2$

Find $\dfrac{dy}{dx}$

9) $5x^3 = (5y^3 + 1)^2$

Find $\dfrac{dy}{dx}$

10) $(y^3 + 2)^2 = 4x^3$

Find $\dfrac{dy}{dx}$

11) $5x^2 + 3 = \cot 4y^2$

Find $\dfrac{dy}{dx}$

12) $5x^3 + 2 = \sin 5y^2$

Find $\dfrac{dy}{dx}$

13) $e^{4y^3} = x^2 + 5$

Find $\dfrac{dy}{dx}$

14) $\ln y^2 = 3x^2 + 1$

Find $\dfrac{dy}{dx}$

15) Find the equation of the tangent line of
$x + 2y^2 = 4$
at the point $(2, 1)$

16) Find the equation of the tangent line of
$4x^3 + 2y^2 = 4$
at the point $(-1, 2)$

17) Find the equation of the tangent line of
$4x^3 y = 5x^3 - 3y^3$
at the point $(-2, 2)$

18) Find the equation of the tangent line of
$3x^2 = 2x^2 y^2 + 4y^3$
at the point $(2, 1)$

19) $-3y^2 + 2 = 2x$

Find $\dfrac{d^2 y}{dx^2}$

20) $4x^3 = 5y^2 + 5$

Find $\dfrac{d^2 y}{dx^2}$

21) $1 = 5x^2 - 2y^2$

Find $\dfrac{d^2 y}{dx^2}$

22) $3x^2 = 5y^2 + 2$

Find $\dfrac{d^2 y}{dx^2}$

23) $4 = 4x - 4y^2$

Find $\dfrac{d^2y}{dx^2}$ at $(2, 1)$

24) $4x^3 = -y^2 + 5$

Find $\dfrac{d^2y}{dx^2}$ at $(-1, 3)$

Derivatives of Inverse Functions

The inverse of a function is a function that undoes or reverses the original function. If the original function is $f(x)$, then the inverse is denoted $f^{-1}(x)$. For example, let's say $g(x) = 8x^3 - 5$, then $g^{-1}(x) = \dfrac{\sqrt[3]{x+5}}{2}$.

If we want to find the inverse of a function, we could just find the inverse first and then find the derivative of the inverse function. But there is a shortcut formula for the derivative of an inverse function.

Derivative of Inverse Function Theorem: Suppose that $f(x)$ is a function with a well-defined inverse $f^{-1}(x)$ and that $f(a) = b$, then: $[f^{-1}]'(b) = \dfrac{1}{f'(a)}$

Derivative of Inverse Function Theorem (second version): $\dfrac{d}{dx}[f^{-1}(x)] = \dfrac{1}{f'(f^{-1}(x))}$

Ex. 1

Let f and g be inverses. The following table lists values for f, g, and f'. Find $g'(7)$.

x	$f(x)$	$g(x)$	$f'(x)$
3	5	7	$-\frac{4}{5}$
5	4	3	$\frac{3}{2}$

Solution: We need to find $g'(7)$. Here we are told that f and g are inverses. So $g'(7) = [f^{-1}]'(7)$. Notice that the table tells us that $f(3) = 7$. So $[f^{-1}]'(7) = \frac{1}{f'(3)} = \frac{1}{\left(-\frac{4}{5}\right)} = -\frac{5}{4}$.

Therefore, $g'(7) = -\frac{5}{4}$.

Ex. 2 $f(x) = 4x - 5$. Find $(f^{-1})'(3)$.

We need to find when $f(a) = 3$. $f(2) = 4(2) - 5 = 3$. We also can find $f'(x) = 4$.

Using the theorem, we have $(f^{-1})'(3) = \frac{1}{f'(2)} = \frac{1}{4}$.

Ex. 3 $f(x) = \sqrt{5x + 2}$. Find $(f^{-1})'(3)$

Solution: We need to find when $f(a) = 3$. Solve $\sqrt{5x + 2} = 3$, gives $x = \frac{7}{5}$. We also find

$f'(x) = \frac{5}{2}(5x + 2)^{-\frac{1}{2}}$. Then, $f'\left(\frac{7}{5}\right) = \frac{5}{2}\left(5\left(\frac{7}{5}\right) + 2\right)^{-\frac{1}{2}} = \frac{5}{2}(7 + 2)^{-\frac{1}{2}} = \frac{5}{2}(9)^{-\frac{1}{2}} = \frac{5}{2} \cdot \frac{1}{3} = \frac{5}{6}$.

Taking the *reciprocal of this, gives* $(f^{-1})'(3) = \frac{1}{f'\left(\frac{7}{5}\right)} = \frac{6}{5}$.

Ex. 4 $f(x) = \sqrt[3]{-4x+3}$. Find $(f^{-1})'(x)$

Solution: Solve for the inverse

$x = \sqrt[3]{-4y+3} \Rightarrow x^3 = -4y+3 \Rightarrow x^3 - 3 = -4y \Rightarrow \frac{x^3-3}{-4} = y$

The inverse is $f^{-1}(x) = \frac{x^3-3}{-4}$. Then find the derivative $f'(x) = \frac{-4}{3}(-4x+3)^{-\frac{2}{3}}$. We will use the second form of the inverse theorem: $\frac{d}{dx}[f^{-1}(x)] = \frac{1}{f'(f^{-1}(x))}$. It's easiest if we do the

composition $f'(f^{-1}(x)) = \frac{-4}{3}\left(-4 \cdot \frac{x^3-3}{-4} + 3\right)^{-\frac{2}{3}} = \frac{-4}{3}(x^3)^{-\frac{2}{3}} = \frac{-4}{3} \cdot \frac{1}{x^2} = \frac{-4}{3x^2}$. Now take the

reciprocal $\frac{1}{f'(f^{-1}(x))} = \frac{1}{\frac{-4}{3x^2}} = -\frac{3x^2}{4}$.

Practice 17

1) Let f and g be inverses.
   The following table list values for f, g and f'

   [x   f(x)   g(x)   f'(x)]
   [4   2   3   -5   ]
   [2   3   4   6   ]

   Find $g'(2)$

2) Let f and g be inverses.
   The following table list values for f, g and f'

   [x   f(x)   g(x)   f'(x)]
   [-2   3   2 $\frac{3}{2}$]
   [3   1   -2 $-\frac{1}{2}$]

   Find $g'(3)$

3) Let f and g be inverses.
   The following table list values for f, g and f'

x	$f(x)$	$g(x)$	$f'(x)$
-1	3	1	$\frac{3}{2}$
3	1	-1	$-\frac{1}{3}$

4) Let f and g be inverses.
   The following table list values for f, g and f'

x	$f(x)$	$g(x)$	$f'(x)$
4	3	5	$-\frac{2}{3}$
5	4	3	$\frac{5}{3}$

5) $f(x) = 2x - 4$
   Find $(f^{-1})'(-1)$

6) $f(x) = 4x + 3$
   Find $(f^{-1})'(0)$

7) $f(x) = -5x + 3$
   Find $(f^{-1})'(2)$

8) $f(x) = x + 2$
   Find $(f^{-1})'(3)$

9) $f(x) = -3x^3 + 4$
   Find $(f^{-1})'(1)$

10) $f(x) = 3x^3 + 5$
    Find $(f^{-1})'(29)$

11) $f(x) = \sqrt[3]{3x+5}$
    Find $(f^{-1})'(3)$

12) $f(x) = \sqrt[3]{5x-4}$
    Find $(f^{-1})'(3)$

13) $f(x) = \sqrt{4x+2}$
    Find $(f^{-1})'(1)$

14) $f(x) = \sqrt{-5x+2}$
    Find $(f^{-1})'(3)$

15) $f(x) = \sqrt{-4x+2}$
    Find $(f^{-1})'(x)$

16) $f(x) = \sqrt{5x-3}$
    Find $(f^{-1})'(x)$

17) $f(x) = \sqrt[3]{5x-5}$
    Find $(f^{-1})'(x)$

18) $f(x) = \sqrt[3]{-4x-1}$
    Find $(f^{-1})'(x)$

19) $f(x) = 3x^2 + 5$, $x \geq 0$
    Find $(f^{-1})'(x)$

20) $f(x) = 2x^3 + 2$
    Find $(f^{-1})'(x)$

Derivatives of Inverse Trigonometric Functions

Inverse trigonometric functions are functions that will undo or reverse the original function. They follow the same properties previously discussed for inverse functions. For trig functions the input is an angle (in radians) and the output is a ratio of sides. For the inverse trig functions the input is the side ratio and the output is the angle. For example if $\sin\frac{\pi}{6} = \frac{1}{2}$ then we can write $\sin^{-1}\frac{1}{2} = \frac{\pi}{6}$ or as $\arcsin\frac{1}{2} = \frac{\pi}{6}$. Notice that there are two ways we can write inverse trig functions, either with a negative one exponent or the *arc* prefix. **Warning** inverse trig functions should not be confused with the reciprocal function: $\sin^{-1} x \neq \frac{1}{\sin x}$.

$\frac{d}{dx}\sin^{-1} u = \frac{1}{\sqrt{1-u^2}}\frac{du}{dx}$	$\frac{d}{dx}\cos^{-1} u = -\frac{1}{\sqrt{1-u^2}}\frac{du}{dx}$
$\frac{d}{dx}\tan^{-1} u = \frac{1}{1+u^2}\frac{du}{dx}$	$\frac{d}{dx}\cot^{-1} u = -\frac{1}{1+u^2}\frac{du}{dx}$
$\frac{d}{dx}\sec^{-1} u = \frac{1}{|u|\sqrt{u^2-1}}\frac{du}{dx}$	$\frac{d}{dx}\csc^{-1} u = -\frac{1}{|u|\sqrt{u^2-1}}\frac{du}{dx}$

Ex. 1 $y = \sin^{-1} 2x^3$ Find $\frac{dy}{dx}$

Solution: We want to find the derivative of inverse sine or arcsine. We will use the formula $\frac{d}{dx}\sin^{-1} u = \frac{1}{\sqrt{1-u^2}}\frac{du}{dx}$ The u for our chain rule is $2x^3$. This gives $\frac{d}{dx}\sin^{-1} 2x^3 =$

$\frac{1}{\sqrt{1-(2x^3)^2}}\frac{d}{dx}(2x^3) = \frac{1}{\sqrt{1-4x^6}}6x^2 = \frac{6x^2}{\sqrt{1-4x^6}}$.

Ex. 2 $f(x) = \text{arcsec } 5x^4$ Find $f'(x)$

Solution: We want to find the derivative of inverse secant or arcsecant. We will use the formula

$\frac{d}{dx}\sec^{-1} u = \frac{1}{|u|\sqrt{u^2-1}}\frac{du}{dx}$ The u for our chain rule is $5x^4$. This gives $\frac{d}{dx}\sec^{-1} 5x^4 =$

$\frac{1}{|5x^4|\sqrt{(5x^4)^2-1}}\frac{d}{dx}(5x^4)$. Notice the absolute value bars in the denominator. If the expression inside can be guaranteed to be positive, you are allowed to drop the absolute value bars. For example, here since the coefficient 5 is positive and the x is to an even exponent, we know that $5x^4$ will always be positive, so we can drop the absolute value bars. $\frac{1}{5x^4\sqrt{25x^8-1}}20x^3 =$

$\frac{20x^3}{5x^4\sqrt{25x^8-1}} = \frac{4}{x\sqrt{25x^8-1}}$. Therefore, $f'(x) = \frac{4}{x\sqrt{25x^8-1}}$

Practice 18

1) $f(x) = \sin^{-1} x^2$

2) $f(x) = \sin^{-1} 3x^5$

3) $f(x) = \cos^{-1} 4x^3$

4) $f(x) = \cos^{-1} 5x^5$

5) $f(x) = \tan^{-1} 5x^5$

6) $f(x) = \tan^{-1} 5x^3$

7) $f(x) = \cot^{-1} x^3$

8) $f(x) = \cot^{-1} 3x^3$

9) $f(x) = \sec^{-1} 3x^2$

10) $f(x) = \sec^{-1} 2x^3$

11) $f(x) = \csc^{-1} -2x^5$

12) $f(x) = \csc^{-1} x^2$

13) $f(x) = \left(\cot^{-1} 4x^3\right)^5$

14) $f(x) = \tan^{-1}\left(-4x^3 + 5\right)^5$

15) $f(x) = \left(\sec^{-1} -2x^2\right)^4$

16) $f(x) = \cot^{-1}\left(4x^3 + 1\right)^5$

17) $f(x) = \arcsin \dfrac{x}{2}$
    Find $f'(1)$

18) $f(x) = \arcsin \dfrac{x}{4}$
    Find $f'(3)$

19) $f(x) = \arccos \dfrac{x}{4}$

Find $f'(3)$

20) $f(x) = \arccos \dfrac{x}{5}$

Find $f'(3)$

21) $f(x) = \arctan \dfrac{x}{5}$

Find $f'(2)$

22) $f(x) = \arctan \dfrac{x}{6}$

Find $f'(8)$

23) $f(x) = \text{arccsc} \dfrac{x}{6}$

Find $f'(9)$

24) $f(x) = \text{arcsec} \dfrac{x}{4}$

Find $f'(6)$

25) $f(x) = \text{arccot} \dfrac{x}{4}$

Find $f'(3)$

26) $f(x) = \text{arccsc} \dfrac{x}{2}$

Find $f'(3)$

27) $f(x) = \text{arccot}\,\dfrac{x}{4}$

Find $f'(3)$

28) $f(x) = \text{arccot}\,\dfrac{x}{4}$

Find $f'(3)$

Derivatives of Higher Order

For many functions we can take the derivative of a function multiple times. The derivative of the derivative is called the second derivative. Likewise, we can have third, fourth, fifth and even higher derivatives. The second derivative is usually denoted by $f''(x)$ or $\dfrac{d^2y}{dx^2}$, the third by $f'''(x)$ or $\dfrac{d^3y}{dx^3}$, and the fourth by $f^{(4)}(x)$ or $\dfrac{d^4y}{dx^4}$.

Ex. 1 $f(x) = x^4 + 3x^3 - x^2$ Find f'''

Solution: We want to find the third derivative. The first derivative is $f'(x) = 4x^3 + 9x^2 - 2x$. The second derivative is the derivative of the first derivative: $f''(x) = 12x^2 + 18x - 2$. The third derivative is the derivative of the second: $f'''(x) = 24x + 18$

Ex. 2 $f(x) = \sec x$ Find f'''

Solution: The first derivative of $\sec x$ is $\tan x \sec x$. $f'(x) = \tan x \sec x$. To find the second derivative we need to use the product rule: $\tan x\,(\tan x \sec x) + \sec x\,(\sec^2 x) =$ $\tan^2 x \sec x + \sec^3 x$. $f''(x) = \tan^2 x \sec x + \sec^3 x$. Now to find the third derivative, we will have to use the product rule, chain rule, and sum rule. $f'''(x) = \tan^2(\tan x \sec x) +$

$\sec x (2 \tan x (\sec^2 x)) + 3 \sec^2 x (\tan x \sec x) = \tan^3 x \sec x + 2 \tan x \sec^3 x +$

$3 \tan x \sec^3 x = \tan^3 x \sec x + 5 \tan x \sec^3 x$. Therefore, $f''' = \tan^3 x \sec x + 5 \tan x \sec^3 x$.

Ex. 3 $f(x) = \sin x$ Find $f^{(125)}$ and $f^{(140)}$

Solution: Here we have to find the 125th derivative which might seem difficult to keep taking the derivative over and over again. Let's take the first few derivatives of sine.

$f(x) = \sin x$

$f'(x) = \cos x$

$f''(x) = -\sin x$

$f'''(x) = -\cos x$

$f^{(4)}(x) = \sin x$

We can see a pattern in the derivatives which has a pattern length of 4. So, to find higher derivatives we can just use the pattern. Divide the order of the derivative by the length of the pattern, and use the remainder to determine where it ends up in the pattern. Here we divide 125 by 4 is 31 with a remainder of 1. The remainder of 1 tells us that $f^{(125)} = f'(x) = \cos x$.

140 divided by 4 gives a remainder of 0, remainder of zero tells us the derivative will be the same as the original function $f^{(140)} = f(x) = \sin x$.

Practice 19

1) $y = -2x^5 - 3x^4 - 4x^3$
   Find $\dfrac{d^4y}{dx^4}$

2) $y = -4x^5 - x^4 + 3x$
   Find $\dfrac{d^4y}{dx^4}$

3) $f(x) = \sqrt{x}$
   Find $f^{(4)}(x)$

4) $f(x) = \sqrt[3]{x}$
   Find $f^{(4)}(x)$

5) $f(x) = \sin 2x$
   Find $f^{(5)}(x)$

6) $f(x) = \cos 3x$
   Find $f^{(5)}(x)$

7) $f(x) = \sec x$
   Find $f''''(x)$

8) $f(x) = \csc x$
   Find $f''''(x)$

9) $f(x) = \cot x$
   Find $f''''(x)$

10) $f(x) = \tan x$
    Find $f''''(x)$

11) $f(x) = e^{3x}$
    Find $f^{(4)}(x)$

12) $f(x) = e^{2x}$
    Find $f^{(5)}(x)$

13) $f(x) = \ln 2x$
    Find $f^{(4)}(x)$

14) $f(x) = \ln 3x$
    Find $f^{(5)}(x)$

15) $f(x) = \sin^{-1} 3x$
    Find $f''(x)$

16) $f(x) = \cos^{-1} 4x$
    Find $f''(x)$

17) $f(x) = \cot^{-1} 2x$
    Find $f''(x)$

18) $f(x) = \tan^{-1} 3x$
    Find $f''(x)$

19) $2x^3 + y^2 = 3$

Find $\dfrac{d^2y}{dx^2}$ in terms of x and y.

20) $5x + 4y^2 = 5$

Find $\dfrac{d^2y}{dx^2}$ in terms of x and y.

21) $1 = 2x^3 - y^2$

Find $\dfrac{d^2y}{dx^2}$ in terms of x and y.

22) $2x^3 - 2y^2 = 3$

Find $\dfrac{d^2y}{dx^2}$ in terms of x and y.

23) $f(x) = \sin x$

Find $f'(x)$

Find $f''(x)$

Find $f'''(x)$

Find $f^{(4)}(x)$

Find $f^{(67)}(x)$

Find $f^{(105)}(x)$

24) $f(x) = \cos x$

Find $\dfrac{dy}{dx}$

Find $\dfrac{d^2y}{dx^2}$

Find $\dfrac{d^3y}{dx^3}$

Find $\dfrac{d^4y}{dx^4}$

Find $\dfrac{d^{55}y}{dx^{55}}$

Find $\dfrac{d^{110}y}{dx^{110}}$

Contextual Applications of Differentiation

Now that we have discussed how to take the derivative of almost any function, or at least any function you will run across in basic calculus, we will start to apply the derivative in various contexts. We will discuss what the derivative actually tells us and how to use the derivative to solve real-world applications.

Interpreting the Meaning of the Derivative in Context

The derivative tells us the instantaneous rate of change of a function at a particular point (usually a specific point in time). Graphically, the derivative tells us the slope of a tangent line at a specific point on the function. Make sure to always include appropriate units and correct interpretation of rate. Also, it is important to track whether the quantity is increasing, positive rate or positive derivative or if the quantity is decreasing with a negative derivative.

Ex. 1 Anthony is riding his hoverboard. The function V gives Anthony's velocity (in meters per second), t seconds after he started riding. What is the best interpretation for the following statement: The value of the derivative of V at $t = 8$ is equal to 2.7.

Solution: The derivative gives us the rate of change of velocity. The rate of change of velocity is also called the acceleration. Therefore, we write: *After 8 seconds, Anthony's acceleration is 2.7 meters per second squared.*

Ex. 2 Maya has taken an initial dose of a prescription medication. The function M gives the amount of medication, in milligrams, in Maya's bloodstream after t hours. What is the best interpretation for the following statement? $M'(6) = -2.5$.

Solution: The derivative gives us the rate of change of medication. We are told that the derivative is a negative value so the amount medication must be decreasing. Therefore, we write: *After 6 hours, the amount of medication is decreasing at a rate of 2.5 milligrams per hour.*

Ex. 3 Julie travels to Alaska. The function D gives the day length, in minutes, of the *nth* day of the year in Anchorage, Alaska. What is the best interpretation for the following statement?

$$\left.\frac{dD}{dn}\right|_{n=18} = 5$$

Solution: The derivative gives us the rate of change of day length. The notation tells us that the derivative at *n = 18* is equal to 5. Therefore, we write: *On January 18, the day length is increasing at a rate of 5 minutes per day.*

GRAPHING CALCULATOR: DERIVATIVE AT A POINT

We have discussed how to take the derivative of a function and then how to evaluate the derivative at a point. Now we will show how to use the calculator to give the value of a derivative at a point.

Press the 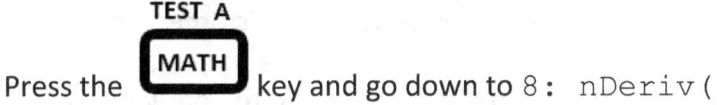 key and go down to `8: nDeriv(`

The screen should look like this and you can type in your variable, function, and value where you want the derivative. For example, if we want the derivative of $f(x) = x^2$ at $x = 4$ we type in everything and press [ENTER]. We get an answer of 8. So $f'(4) = 8$.

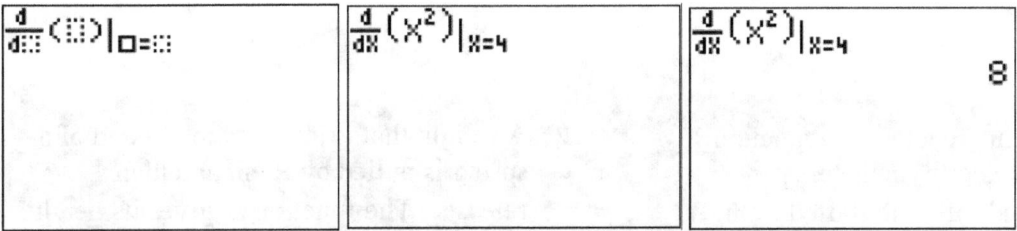

If you have an older calculator your screen may look like this:

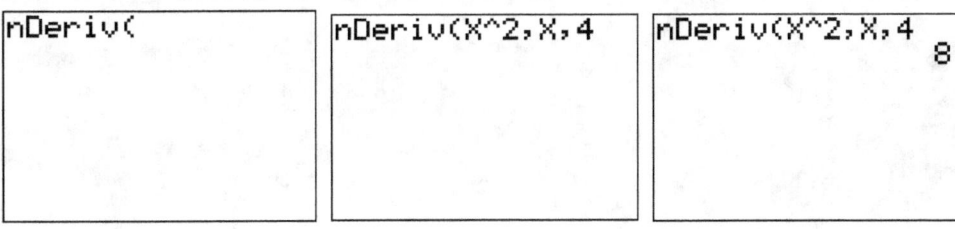

Ex. 4 Michael uploaded a funny video on his website, which rapidly gains views over time. The following function gives the number of views t days after Michael uploaded the video: $V(t) = 120e^{.4t}$. What is the instantaneous rate of change of the number of views 6 days after the video was uploaded?

Solution: When we need to find the instantaneous rate of change at a specific time, that is telling us we need to find the derivative at that value. Here we need to find the derivative of the given function at $t = 6$. We can do this by hand or with the calculator. If we use the calculator, make sure to change the variable to x instead of t.

$$\frac{d}{dx}(120e^{.4x})|_{x=6}$$
$$529.1124804$$

We need to round this value, since we can only have a whole number of views. This gives us the solution: *529 views/day*.

Practice 20

1) Ethan is riding his scooter. The function V gives Ethan's velocity (in meters per second), t seconds after he started riding. What is the best interpretation for the following statement?
The value of the derivative of V at $t = 6$ is equal to 2.5

2) A weight that is attached to the end of a spring is pulled by Ryan and then released. The function H gives its height, in centimeters, after t seconds. What is the best interpretation for the following statement?
$H'(0) = 4$

3) Gretchen fills her pool with water. The function W gives the pool's water level (in meters) after t minutes. What is the best interpretation for the following statement? The slope of the line tangent to the graph of W at $t = 6$ is equal to 0.2.

4) Lily's goldfish pond is leaking. The function W gives the pond's water level (in centimeters) after h hours. What is the best interpretation for the following statement?
The slope of the line tangent to the graph of W at $t = 4$ is equal to -3.

5) Bryan throws a football. The function v gives the ball's velocity, in meters per second, after t seconds. What is the best interpretation for the following statement?
$v'(0) = -10$

6) Cara has taken an initial dose of a prescription medication. The function M gives the amount of medication, in milligrams, in Cara's bloodstream after t hours. What is the best interpretation for the following statement?
$M'(4) = -1.5$

7) Natalie travels to Alaska. The function D gives the day length, in minutes, of the nth day of the year in Anchorage, Alaska. What is the best interpretation for the following statement?
$\left.\dfrac{dD}{dn}\right|_{n=12} = 4$

8) The function s gives the total distance, Matthew has driven (in miles) after t hours.
What is the best interpretation for the following statement?
$\left.\dfrac{ds}{dt}\right|_{t=2} = 70$

9) The function s gives the total distance, Katie has driven (in kilometer) after t hours.
What is the best interpretation for the following statement?
$\left.\dfrac{ds}{dt}\right|_{t=5} = 90$

10) Ashanti decides to start saving for retirement. A function I gives the value of a retirement investment after y years
What is the best interpretation of
$\left.\dfrac{dI}{dy}\right|_{y=10} = 2000$

11) Brandon uploaded a funny video on his website, which rapidly gains views over time. The following function gives the number of views t days after Brandon uploaded the video:
$V(t) = 100e^{0.3t}$
What is the instantaneous rate of change of the number of views 5 days after the video was uploaded?

12) The following function gives the cost, in dollars, of producing x kilograms of fertilizer:
$C(x) = 0.001x^3 - 0.14x^2 + 7x + 160$
What is the instantaneous rate of change of the cost when 100 kilograms are produced?

13) The following function gives the temperature (in degrees Celsius) at the beach in Miami, Florida, t hours after midnight on a certain day:
$M(t) = -6 \sin \dfrac{\pi t}{12} + 18$
What is the instantaneous rate of change of the temperature at 10 a.m.?

14) A water tank is drained. The following function gives the volume, in liters, of the water remaining in the tank t minutes after the drain is opened:
$V(t) = 3000(1 - 0.05t)^2$

What is the instantaneous rate of change of the volume after 12 minutes?

Motion Problems

Calculus was developed to solve problems in the real world involving changing quantities. One such type of problem has to do with motion of objects. We will be dealing with motion in one dimension, either left/right or up/down. These problems involve analyzing the motion of a particle with respect to its position, direction, velocity, and acceleration. Position is usually given as a function and may be written as $s(t)$, or $x(t)$ if it is moving along the x-axis or $y(t)$ if it is moving along the y-axis.

Velocity is the rate of change of the position and will be given by the derivative of the position function. Graphically, velocity will be the slope of a tangent on the graph of the position function. Velocity is a vector which means it has a magnitude (numerical value) and a direction. A positive velocity indicates the particle is moving right (or up) while a negative velocity indicates the particle is moving left (or down). Speed is just the magnitude of the velocity without any indication of the direction. So, speed is given by the absolute value of the velocity.

Acceleration is the rate of change of the velocity and will be given by the derivative of the velocity function. This will be the second derivative of the position function. Graphically, acceleration will be the slope of a tangent on the graph of the velocity function. Acceleration is also considered a vector, so the sign of acceleration is important. If the acceleration is positive that means the velocity is increasing. A negative acceleration means the velocity is decreasing. A particle is speeding up, or accelerating, when the velocity and acceleration are acting in the same direction, that is acceleration and velocity have the same sign. A particle is slowing down, or decelerating, when the velocity and acceleration are acting in opposite directions, that is acceleration and velocity have opposite signs.

Those who have studied physics may also be familiar between the difference between distance and displacement. Displacement is just how far we have traveled from where we started. Displacement only cares about the start and the end, not what happened in the middle. Displacement is the difference between the final position and the initial (or starting) position. Displacement may be positive, meaning the particle ended up to the right (or above) the starting position. Or displacement, may be negative, meaning the particle ended up to the left (or below) its starting position. On the other hand, *total distance* is the sum of all distances traveled. Distances will always be positive (unless the particle has not moved at all and has a distance of 0).

SUMMARY:
Position = $s(t)$, $x(t)$, or $y(t)$
Velocity = $v(t) = s'(x)$
   Velocity = Slope on a position graph
   $v > 0$ **the particle is moving right (up), and s is increasing**
   $v < 0$ **the particle is moving left (down), and s is decreasing**
   Speed is $|v(t)|$, the absolute value of velocity, therefore always nonnegative
Acceleration= $a(t) = v'(x) = s''(x)$
   Acceleration = Slope on a velocity graph
   If $a > 0$, then velocity is increasing
   If $a < 0$, then velocity is decreasing
   If a and v are the same sign, paarticle is speeding up (accelerating)
   If a and v are opposite signs, particle is slowing down(decelerating)
   The particle reverses (changes) direction when $v = 0$ and $a \neq 0$
Displacement = Final position - intitial position $\Delta s = s_f - s_0$

Interpreting Motion Graphs

Ex. 1 Here we will analyze the graph below. The graph gives the motion of a particle moving horizontally for 10 seconds. Determine when the particle is moving forward and when the particle is moving backward.

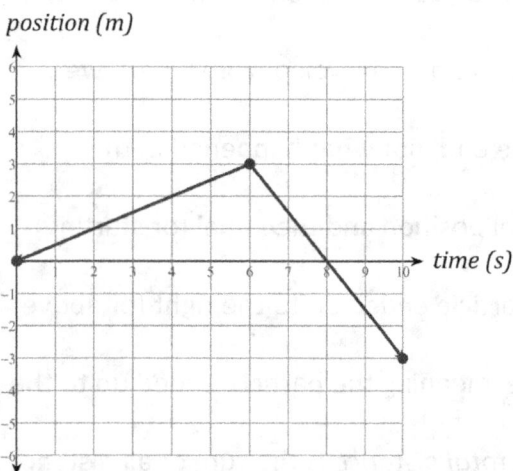

Solution: This graph gives the position (in meters) of a particle over time (in seconds). The graph will be moving forward when it has a positive velocity. It will be moving backwards when it has a negative velocity. We can find the velocity from this graph by looking at the slopes of the lines. The slope of the left side from t = 0 to t = 6, has a slope of 1/2. The slope on the right side from t = 6 to t = 10 is – 2. Therefore, the particle is moving forward on the interval (0, 6). The particle is moving backward on the interval (6, 10). Notice that at t = 6, the particle switched directions.

Ex. 2 Here we will analyze the graph below. The graph gives the velocity of a particle moving horizontally for 10 seconds. Determine when the particle is speeding up and when the particle is slowing down.

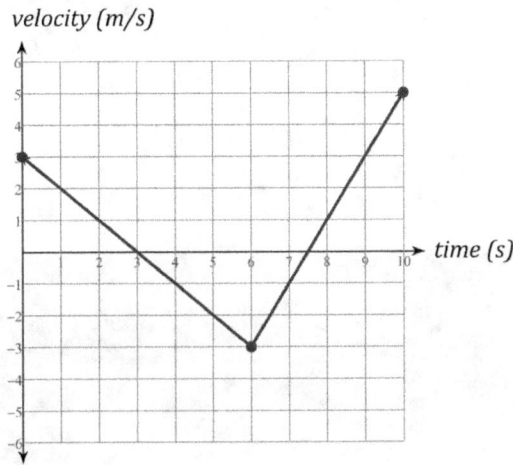

Solution: This graph gives the velocity (in meters per second) of a particle over time (in seconds). Acceleration is given by the slope on a velocity/time graph. The particle will be speeding up when the signs of the velocity and acceleration are the same. That is when the either: 1) the function is above the x-axis and the slope is positive or 2) the function is below the x-axis and the slope is negative. The particle will be slowing down when the signs of the velocity and acceleration are the opposite of each other. That is when the either: 1) the function is above the x-axis and the slope is negative or 2) the function is below the x-axis and the slope is positive. Therefore, the particle is speeding up on the intervals (3, 6) and (7.5, 10). The particle is slowing down on the intervals (0, 3) and (6, 7.5). Notice that the particle is at rest at times t = 3 and t = 7.5.

Ex. 3 The graph gives the position of a particle moving horizontally for 20 seconds. Answer the questions below:

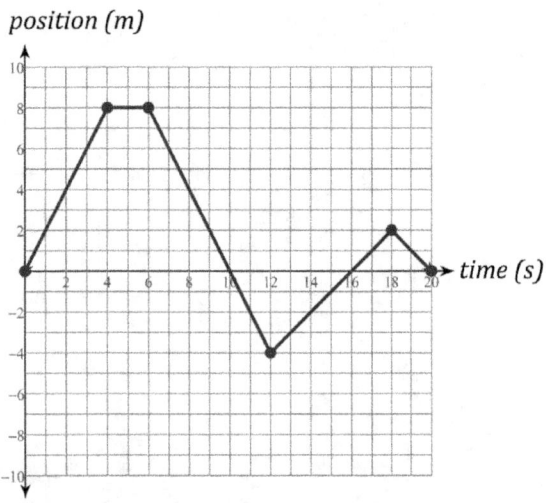

What is the position at t = 3? *Find the point at t = 3 on the graph. The position is 6 m*

What is the velocity at t = 3? *Find the slope at t = 3. The velocity is 2 m/s*

What is the position at t = 14? *The position is –2 m*

What is the velocity at t = 14? *The velocity is 1 m/s*

What is the speed at t = 2? *Speed is the absolute value of velocity. The speed is 2 m/s*

What is the speed at t = 8? *The velocity is – 2 m/s. Therefore, speed is 2 m/s*

When is the particle moving right? *The particle is moving to the right when velocity is positive, when the slope is positive. The particle is moving to the right on intervals (0, 4) and (12, 18)*

When is the particle moving left? *The particle is moving to the left when velocity is negative, when the slope is negative. The particle is moving to the left on intervals (6, 12) and (18, 20)*

When is the particle at rest? *The particle is at rest when the velocity is zero, or when the particle switches direction. The particle is at rest from t = 4 to t = 6. It is also at zero for an instant at t = 12 and t = 18.*

When does the particle change direction? *The particle changes direction when the slope changes sign. The particle changes direction at t = 6, t = 12, t = 18*

What is the displacement? *Displacement is the final position minus the initial position. Since the particle ends at 0 and begins at 0, the displacement is 0.*

What is the total distance traveled? *To find the total distance, we need to add up all the distances traveled. In the first 6 seconds it travels 8 m, then by t = 12 it travels 12 m, then travels 6 m, finally traveled 2 m. 8 + 12 + 6 + 2= 28 m*

Ex. 4 The graph gives the velocity of a particle moving vertically for 20 seconds. Answer the questions below:

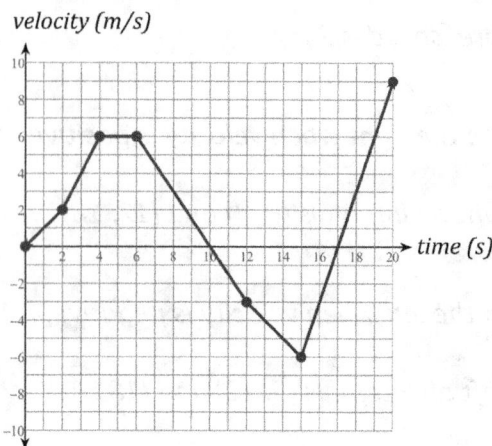

What is the velocity at t = 3? *The velocity at t = 3 is given by its height at that value. 4 m/s*

What is the acceleration at t = 3? *The acceleration is given by the slope. 2 m/s²*

What is the velocity at t = 5? *6 m/s*

What is the acceleration at t = 5? *0 m/s²*

What is the velocity at t = 14? *−5 m/s*

What is the acceleration at t = 14? *−1 m/s²*

What is the speed at t = 8? *Speed is given by the absolute value of the velocity. Since the velocity is 3 m/s, the speed is also 3 m/s*

What is the speed at t = 12? *Since the velocity is −3 m/s, the speed is 3 m/s*

When is the particle moving up? *The particle moves up when velocity is positive. The graph is above the x-axis This happens on the intervals (0, 10) and (17, 20)*

When is the particle moving down? *The particle moves down when velocity is negative. The graph is below the x-axis. This happens on the interval (10, 17)*

When is the particle at rest? *The particle is at rest when the velocity is 0. This happens at t = 0, t = 10, and t = 17*

When does the particle change direction? *The particle changed direction when the sign of the velocity changes. t = 10 and t = 17*

When is the particle speeding up? *The particle is speeding up when the velocity and acceleration have the same sign. This happens on the intervals (0, 4), (10, 15), and (17, 20)*

When is the particle slowing down? *The particle is slowing down when the velocity and acceleration have the opposite signs. This happens on the intervals (6, 10) and (15, 17)*

Practice 21

1.
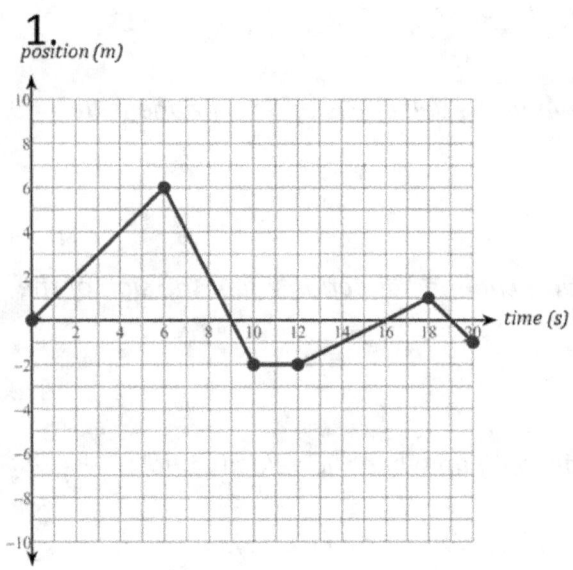

2.
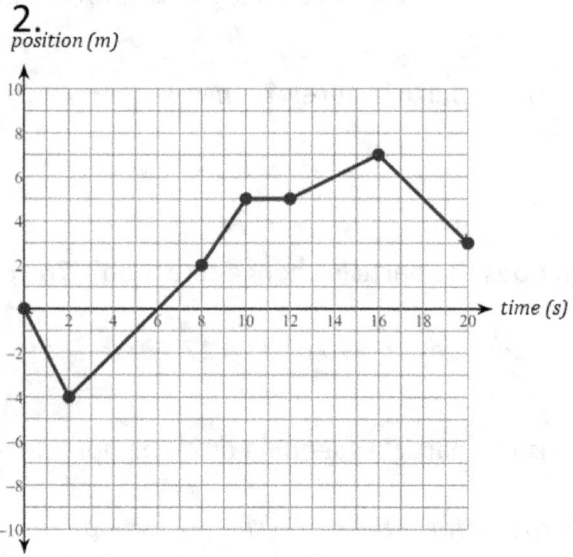

What is the position at $t = 7$?

What is the velocity at $t = 7$?

What is the speed at $t = 7$?

What is the position at $t = 14$?

What is the velocity at $t = 14$?

What is the speed at $t = 14$?

When is the particle moving right?

When is the particle moving left?

When is the particle at rest?

When does the particle change direction?

What is the displacement?

What is the total distance traveled?

What is the position at $t = 4$?

What is the velocity at $t = 4$?

What is the position at $t = 14$?

What is the velocity at $t = 14$?

What is the speed at $t = 4$?

What is the speed at $t = 17$?

When is the particle moving right?

When is the particle moving left?

When is the particle at rest?

When does the particle change direction?

What is the displacement?

What is the total distance traveled?

3.

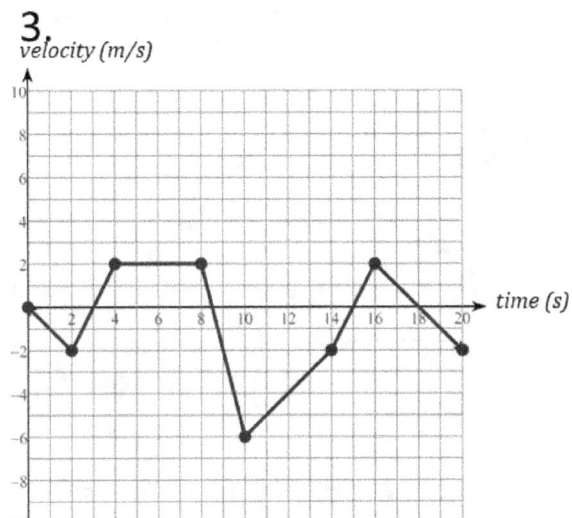

What is the velocity at $t = 3$?

What is the acceleration at $t = 3$?

What is the velocity at $t = 5$?

What is the acceleration at $t = 5$?

What is the velocity at $t = 14$?

What is the acceleration at $t = 14$?

What is the speed at $t = 8$?

What is the speed at $t = 12$?

When is the particle moving right?

When is the particle moving left?

When is the particle at rest?

When does the particle change direction?

When is the particle speeding up?

When is the particle slowing down?

4.

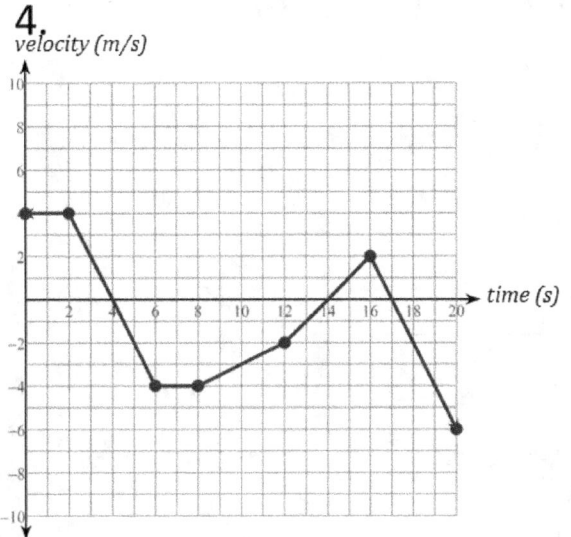

What is the velocity at $t = 4$?

What is the acceleration at $t = 4$?

What is the velocity at $t = 7$?

What is the acceleration at $t = 7$?

What is the velocity at $t = 15$?

What is the acceleration at $t = 15$?

What is the speed at $t = 2$?

What is the speed at $t = 12$?

When is the particle moving right?

When is the particle moving left?

When is the particle at rest?

When does the particle change direction?

When is the particle speeding up?

When is the particle slowing down?

Given the position vs. time graphs below, describe the velocity and acceleration in words.

5)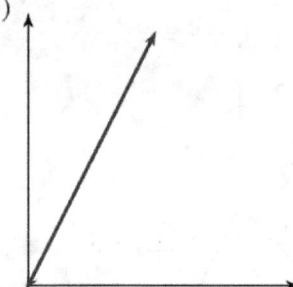

6)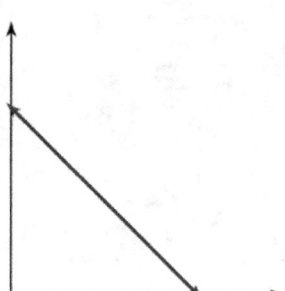

7)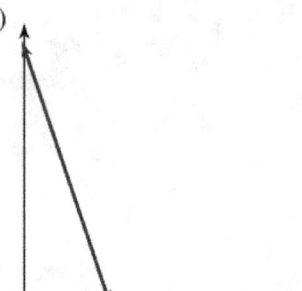

8)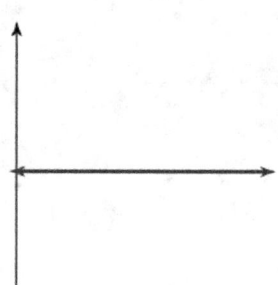

9)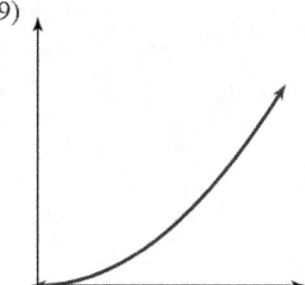

10)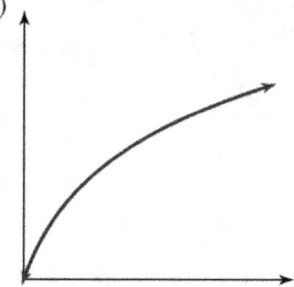

11)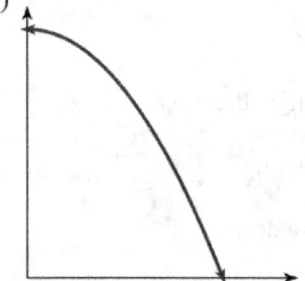

12)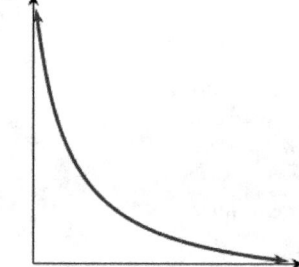

Draw the position vs. time graphs that fit the description.

13) Moving left, slowing down.

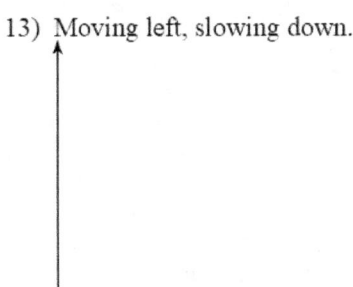

14) Moving right at constant velocity.

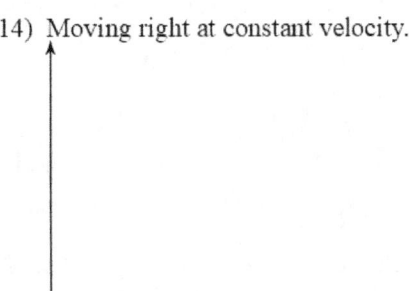

15) Moving right fast, slowing down.

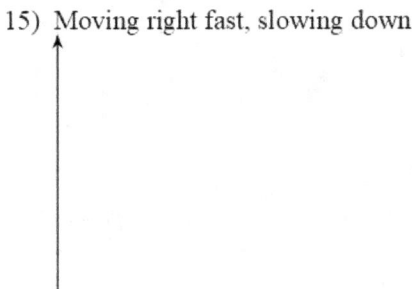

16) Not moving at all

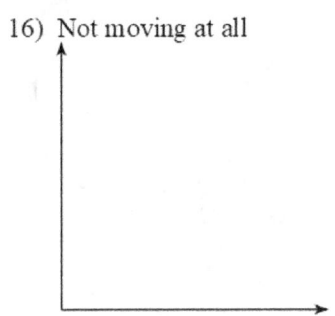

17) Moving left, speeding up

18) Constant negative slow velocity.

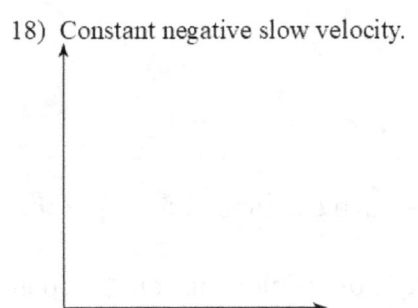

19) Moving right, accelerating.

20) Moving right, then halfway through moving left back to starting position.

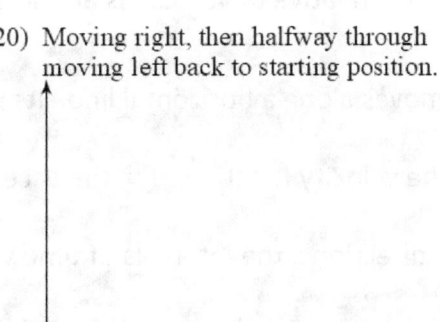

21) Moving left at constant speed, then moving right accelerating.

22) A ball at rest for some time on a building, falls to the ground.

23) You walk away from your house, then you rest, then you run back.

24) A man jumps from a plane, then pulls the parachute.

Motion Problems Using Calculus

Now that we have examined motion problems from graphs, we will use calculus to help us work with functions describing motion. Keep in mind that the derivative of position is velocity, while the derivative of velocity is acceleration.

Ex. 1 A particle moves along a horizontal line. Its position function is $s(t) = t^3 - 18t^2 + 81t$ for $t \geq 0$. Find the velocity function $v(t)$ the acceleration function $a(t)$, the times t when the particle changes directions, the intervals of time when the particle is moving left and moving right, the times t when the acceleration is 0, and the intervals of time when the particle is slowing down and speeding up.

Solution: First, we find the velocity function by taking the derivative: $v(t) = s'(t) = 3t^2 - 36t + 81$. Then the acceleration, is the second derivative: $a(t) = s''(t) = 6t - 36$. The particle will change direction when the velocity is zero and the velocity changes sign. Set the velocity to zero and solve for t: $3t^2 - 36t + 81 = 0 \Rightarrow 3(t^2 - 12t + 27) = 0 \Rightarrow 3(t-9)(t-3) = 0 \Rightarrow t = 3 \text{ or } t = 9$. Then we find where the function is moving left or right by evaluating the velocity function inside each of the intervals. For example, evaluate. $v(2), v(4), v(10)$.

$v(2) = 3(2)^2 - 36(2) + 81 = 21$

$v(4) = 3(4)^2 - 36(4) + 81 = -15$

$v(10) = 3(10)^2 - 36(10) + 81 = 21$

We use these values to make a sign chart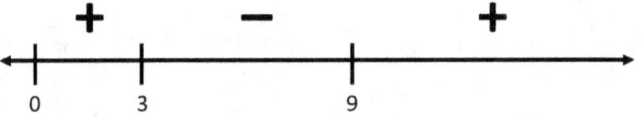

This tells us that the function is moving left on the interval $(3, 9)$ and it moves right on the intervals $[0, 3), (9, \infty)$.

Now we determine where the acceleration is zero: $6t - 36 = 0 \Rightarrow t = 6$. We determine the sign of the acceleration before and after this point by evaluating the acceleration at a value less than 6 and a value greater. $a(2) = 6(2) - 36 = -20$ and $a(20) = 6(10) - 36 = 24$

We expand our sign chart to include the acceleration in addition to the velocity

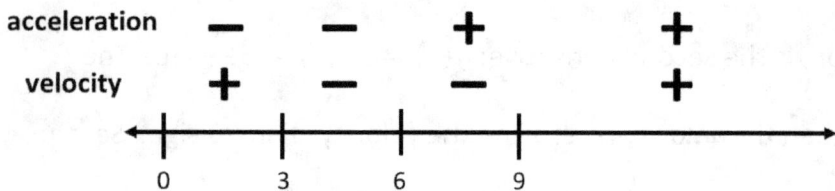

We use the sign chart to determine when the particle is slowing down, that is when the velocity and acceleration have opposite signs. The particle slows down on the intervals: $[0, 3), (6, 9)$.

The particle is speeding up when the velocity and acceleration have the same sign. The particle speeds up on the intervals: $(3, 6), (9, \infty)$.

Ex. 2 A particle moves along a horizontal line. A particle moves along a horizontal line. Its position function is $x(t) = -t^3 + 18t^2 - 81t$ for times $2 \leq t \leq 5$. Find the maximum speed and times [t] when this speed occurs, the displacement of the particle, and the distance traveled by the particle over the given interval.

Solution: The velocity is given by $v(t) = -3t^2 + 36t - 81$. The velocity will have a maximum at its critical points, either at the endpoints or at its turning points. The turning points can occur when the derivative of the velocity is equal to zero. $v'(t) = -6t + 36$ Set to zero: $-6t + 36 = 0 \Rightarrow t = 6$. Time t = 6 is outside the given interval of $2 \leq t \leq 5$. Check at the endpoints: $v(2) = -3(2)^2 + 36(2) - 81 = -21$ or $v(5) = -3(5)^2 + 36(5) - 81 = 24$. Since speed is the absolute value, we have maximum speed of 24 when t = 5. Displacement is found by taking the difference between the final position and the initial position. $x(2) = -(2)^3 + 18(2)^2 - 81(2) = -98$ and $x(5) = -(5)^3 + 18(5)^2 - 81(5) = -80$. The displacement is $-80 - (-98) = 18$. To find the total distance, we need to find out when the particle is changing directions by taking the derivative and setting it to zero. $v(t) = -3t^2 + 36t - 81 = 0 \Rightarrow -3(t^2 - 12t + 27) = 0 \Rightarrow -3(t-9)(t-3) \Rightarrow t = 3 \text{ and } t = 9$. Now we evaluate the position function at the endpoints of the interval and each time (in the given interval) that the velocity is zero:

$$x(2) = -(2)^3 + 18(2)^2 - 81(2) = -98$$

$$x(3) = -(3)^3 + 18(3)^2 - 81(3) = -108$$

$$x(5) = -(5)^3 + 18(5)^2 - 81(5) = -80$$

Now we find the distance between each of these and add together

$$|-108 - (-98)| + |-80 - (-108)| = 38$$

Therefore, the maximum speed is 24 at t = 5; the displacement is 18; the distance traveled is 38.

Practice 22

1) A particle moves along a horizontal line.
   Its position function is
   $s(t) = -t^2 + 9t + 10$
   Find the position at $t = 5$
   Find the velocity at $t = 5$
   Find the speed at $t = 5$
   Find the acceleration at $t = 5$

2) A particle moves along a horizontal line.
   Its position function is
   $s(t) = t^4 - 14t^3$
   Find the position at $t = 8$
   Find the velocity at $t = 8$
   Find the speed at $t = 8$
   Find the acceleration at $t = 8$

3) A particle moves along a horizontal line. Its position function is
$s(t) = t^3 - 22t^2 + 121t$
Find the position at $t = 5$
Find the velocity at $t = 5$
Find the speed at $t = 5$
Find the acceleration at $t = 5$

4) A particle moves along a horizontal line. Its position function is
$s(t) = t^3 - 12t^2$
Find the position at $t = 5$
Find the velocity at $t = 5$
Find the speed at $t = 5$
Find the acceleration at $t = 5$

5) A particle moves along a horizontal line. Its position function is
$s(t) = -t^3 + 22t^2 - 121t$
Find the position at $t = 6$
Find the velocity at $t = 6$
Find the speed at $t = 6$
Find the acceleration at $t = 6$

6) A particle moves along a horizontal line. Its position function is
$s(t) = t^3 - 23t^2 + 120t$
Find the position at $t = 3$
Find the velocity at $t = 3$
Find the speed at $t = 3$
Find the acceleration at $t = 3$

A particle moves along a horizontal line. Its position function is $s(t)$ for $t \geq 0$. For each problem, find the velocity function $v(t)$, the acceleration function $a(t)$, the times t when the particle changes directions, the intervals of time when the particle is moving left and moving right, the times t when the acceleration is 0, and the intervals of time when the particle is slowing down and speeding up.

7) $s(t) = t^2 - 16t + 64$

8) $s(t) = t^2 - 25t + 150$

9) $s(t) = -t^3 + 16t^2 - 64t$

10) $s(t) = t^3 - 13t^2 + 40t$

11) $s(t) = -t^4 + 8t^3$

12) $s(t) = t^4 - 10t^3$

A particle moves along a horizontal line. Its position function is $s(t)$ for $t \geq 0$. For each problem, find the maximum speed and times t when this speed occurs, the displacement of the particle, and the distance traveled by the particle over the given interval.

13) $s(t) = t^2 - 12t + 27$; $4 \leq t \leq 10$

14) $s(t) = -t^2 + 20t - 84$; $9 \leq t \leq 15$

15) $s(t) = -t^3 + 13t^2 - 40t$; $0 \leq t \leq 3$

16) $s(t) = -t^3 + 4t^2 + 60t$; $0 \leq t \leq 10$

17) $s(t) = t^4 - 10t^3$; $2 \leq t \leq 8$

18) $s(t) = -t^4 + 9t^3$; $5 \leq t \leq 8$

19) $s(t) = t^3 - 22t^2 + 105t$; $3 \leq t \leq 7$

20) $s(t) = -t^3 + 14t^2$; $6 \leq t \leq 11$

Related Rates

Related rates are another common calculus problem. In many real world applications, there may be several rates that are related to each other. For example, if you blow up a balloon, the volume increases and the surface area increases, but the rate that the volume increases is not the same as the rate that the surface area increases. Another common example might be water in a tank shaped like a cone. If the rate that the volume is changing is known, then we can find the rate of the change in the height of the water at a particular time. Related rates are generally more difficult calculus problems because they require multiple steps and there are many different types of related rates problems. We will go through the general steps in solving the more common types you will likely come across.

Steps to Solving Related Rates Problems:

1. Draw a picture if possible.

2. Make a list of given variables/rates.

3. Pay attention to units. Rates will be written as derivatives, ex. $\frac{dx}{dt}$.

4. Pay attention to whether a rate is increasing (positive) or decreasing (negative)

2. Make a list of variables to be determined.

3. Write the equation(s) that relates the variables.

4. Implicitly differentiate both sides.

5. Solve for the variable/rate to be determined.

6. Now substitute. Make sure your units make sense.

Ex. 1 A 25 ft ladder is leaning against a wall and sliding towards the floor. The top of the ladder is sliding down the wall at a rate of 2 ft/sec. How fast is the base of the ladder sliding away from the wall when the base of the ladder is 7 ft from the wall?

Solution: First step, draw a picture. It should be noted that you don't need to make it a realistic drawing, just enough detail to label and find the correct formula. A ladder falling against a wall will form a right triangle. Label all the quantities we know. The length of the ladder is 25 ft and is constant, so the rate of change for the length of the ladder will not change. I draw arrows on the diagram to help picture what variables are increasing and decreasing. As the ladder slides down, the vertical distance will decrease and the horizontal distance will increase. This means that the rate $\frac{dy}{dt}$ is negative and rate $\frac{dx}{dt}$ will be positive.

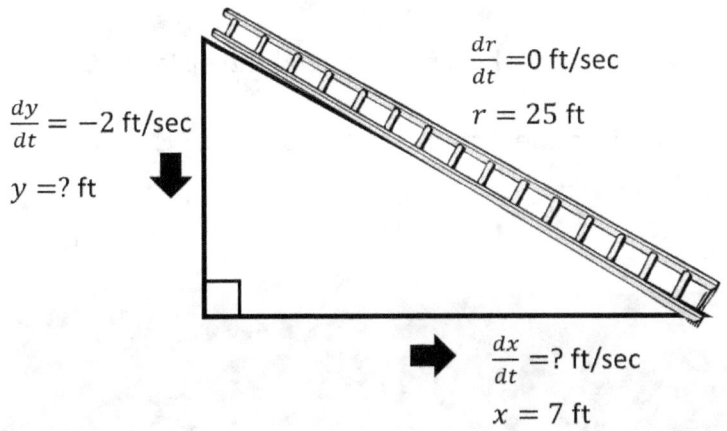

What formula do we need for a right triangle? Pythagorean theorem should jump to mind. $x^2 + y^2 = r^2$. Since r will remain constant, we can substitute 25 in for r. $x^2 + y^2 = 25$ The variables x and y will be changing so we can not substitute until we have taken the derivative. Implicitly differentiate to give: $2x\frac{dx}{dt} + 2y\frac{dy}{dt} = 0$. Now we have an equation with related rates.

We have the known rate $\frac{dy}{dt}$ and we need to find $\frac{dx}{dt}$. We also know $x = 7$ at a particular time, so we need to solve for y at that time. Again, using Pythagorean theorem we have: $7^2 + y^2 = 25^2 \Rightarrow 49 + y^2 = 625 \Rightarrow y^2 = 576 \Rightarrow y = 24$. Take our related rates equation and substitute in what we know: $2x\frac{dx}{dt} + 2y\frac{dy}{dt} = 0 \Rightarrow 2(7)\frac{dx}{dt} + 2(24)(-2) = 0 \Rightarrow 14\frac{dx}{dt} - 96 = 0 \Rightarrow 14\frac{dx}{dt} = 96 \Rightarrow \frac{dx}{dt} = \frac{48}{7}$. Therefore, the ladder is sliding away at a rate of $\frac{48}{7}$ ft/sec.

Ex. 2 A 6 ft tall person is walking towards a 20 ft tall lamppost at a rate of 6 ft/sec. Assume the scenario can be modeled with right triangles. At what rate is the length of the person's shadow changing when the person is 15 ft from the lamppost?

Solution: First step, draw a picture. Label everything we are given. The height of the lamppost, the height of the person, the rate of movement towards the lamppost, and the distance to the lamppost. Notice that the rate will be negative since the distance to the lamppost is decreasing.

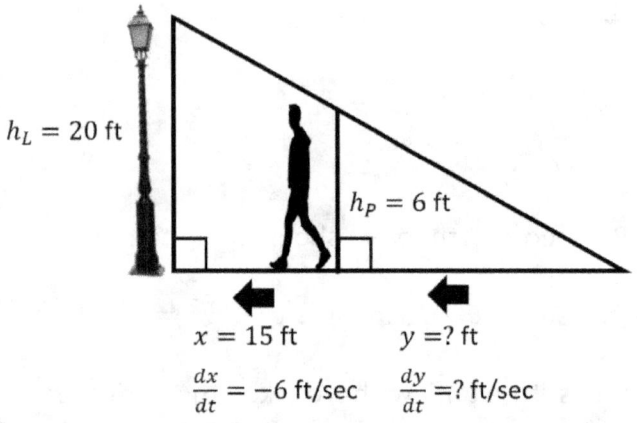

$h_L = 20$ ft

$h_P = 6$ ft

$x = 15$ ft

$y = ?$ ft

$\frac{dx}{dt} = -6$ ft/sec

$\frac{dy}{dt} = ?$ ft/sec

After we draw the picture, it should be clear that this problem will involve similar triangles. We use that fact to set up a formula, a proportion expressing the quantities we know and those we want to find. We set up a proportion involving the bottom side divided by the height for each triangle as: $\frac{x+y}{20} = \frac{y}{6}$. We can simplify this equation a bit before we differentiate to make it easier: $6x + 6y = 20y \Rightarrow 6x = 14y \Rightarrow \frac{3}{7}x = y$. Take the derivative with respect to time to get: $\frac{3}{7}\frac{dx}{dt} = \frac{dy}{dt}$. Now we can substitute $\frac{3}{7}(-6) = \frac{dy}{dt} \Rightarrow \frac{dy}{dt} = -\frac{18}{7}$. Therefore, the shadow's length is decreasing at a rate of $-\frac{18}{7}$ ft/sec.

Ex. 3 Water leaking onto a floor forms a circular pool. The radius of the pool increases at a rate of 5 cm/min. How fast is the area of the pool increasing when the radius is 14 cm?

Solution: First step, draw a picture. We draw a circle with a radius. We label the rate that the radius is increasing, the length of the radius at a certain time. We want to find the rate that the area is changing, so we label the area. I put arrows going out from the circle, to show that the circle is expanding. The rate for the radius and the area should be positive then.

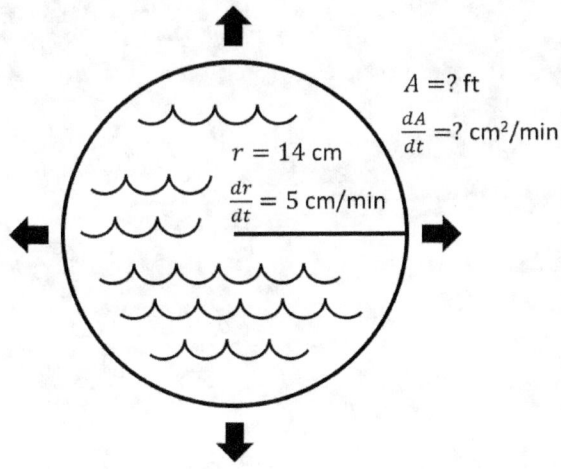

We need a formula to relate area of a circle to the radius of a circle. We use the area formula: $A = \pi r^2$. Since the area and radius are both changing, we leave them as variables and then differentiate implicitly to get: $\frac{dA}{dt} = 2\pi r \frac{dr}{dt}$. Now we can substitute in the values that we know for $\frac{dr}{dt}$ and r: $\frac{dA}{dt} = 2\pi(14)(5) = 140\pi$. Therefore, the area is increasing at a rate of 140π cm² /min.

Ex. 4 A balloon rises at the rate of 8 ft/sec from a point on the ground 60 ft from the observer. How do you find the rate of change of the angle of elevation when the balloon is 25 ft above the ground?

Solution: First step, draw a picture, label what we know, and label what we want to know. This problem involves a triangle and we are given information about two sides of a triangle and we need to find the rate of change for the angle of elevation. We are prompted to use some trigonometry to set up this problem. .

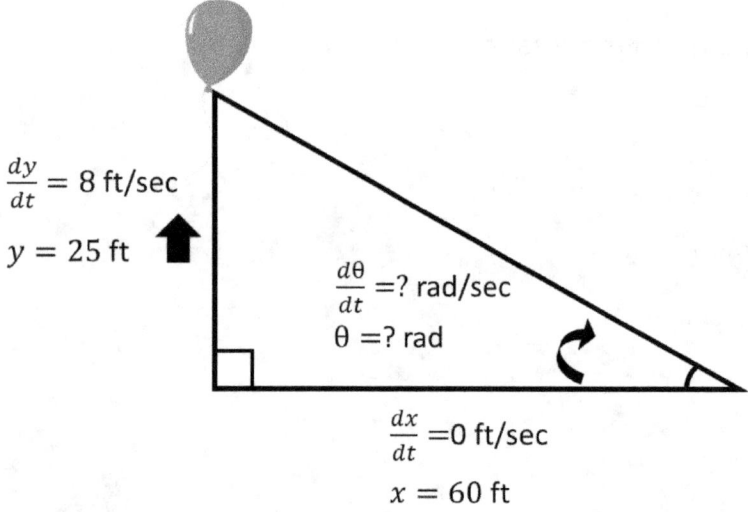

$\frac{dy}{dt} = 8$ ft/sec

$y = 25$ ft

$\frac{d\theta}{dt} = ?$ rad/sec

$\theta = ?$ rad

$\frac{dx}{dt} = 0$ ft/sec

$x = 60$ ft

Since we are given the opposite and adjacent sides corresponding to the angle, we will use tangent: $\tan\theta = \frac{y}{x}$ We may plug in values for variables that are constant into our formula.

Here the distance on the ground, x, is constant which gives us: $\tan\theta = \frac{y}{60} \Rightarrow 60\tan\theta = y$

Differentiate implicitly, gives: $60\sec^2\theta \frac{d\theta}{dt} = \frac{dy}{dt}$ Solve for the rate of the angle: $\frac{d\theta}{dt} = \frac{\cos^2\theta}{60}\frac{dy}{dt}$.

We can find cosine of theta by first getting the hypotenuse $hyp = \sqrt{25^2 + 60^2} = 65$. So, $\cos\theta = \frac{60}{65} = \frac{12}{13}$. Substitute what we know into the derivative formula: $\frac{d\theta}{dt} = \frac{\cos^2\theta}{60}\frac{dy}{dt} = \frac{1}{60}\left(\frac{12}{13}\right)^2(8) = \frac{96}{845}$. Therefore, the angle of elevation is increasing at a rate of $\frac{96}{845}$ rad/sec.

Ex. 5 A conical paper cup is 20 cm tall with a radius of 10 cm. The bottom of the cup is punctured so that the water leaks out at a rate of $\frac{2\pi}{3}$ cm³/sec. At what rate is the water level changing when the water level is 3 cm?

Solution: First step, draw a picture. We draw a cone with a radius. Really all we need is the triangular slice shown to the right. This shows that we are working with similar triangles.

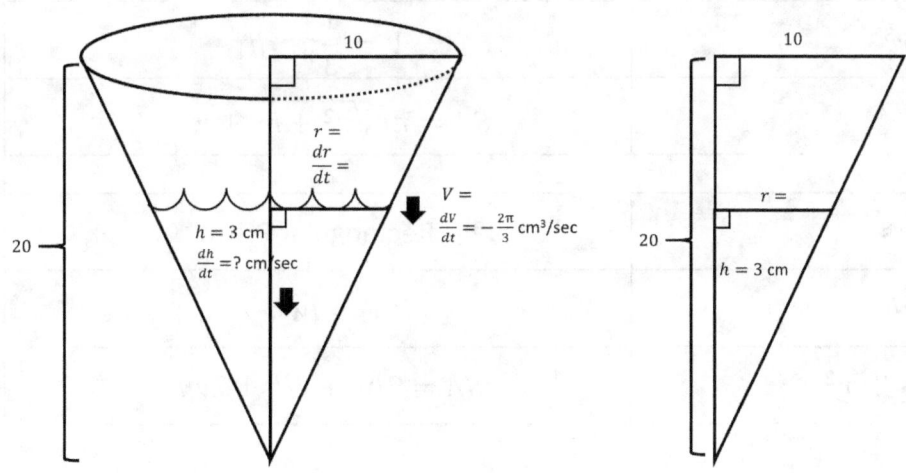

We are given information about the volume of the water and the height of the water level. Since we are not given any information about the radius, we need to express the height in terms of the radius. We can use the fact that we have similar triangles and set up a proportion: $\frac{r}{h} = \frac{10}{20} \Rightarrow r = \frac{1}{2}h$. This means we can substitute in for r. The formula for the volume of a cone is: $V = \frac{1}{3}\pi r^2 h$. After substitution, we get: $V = \frac{1}{3}\pi(\frac{1}{2}h)^2 h = \frac{1}{3}\pi\frac{1}{4}h^3 = \frac{1}{12}\pi h^3$. Take the derivative with respect to time: $\frac{dV}{dt} = \frac{3}{12}\pi h^2 \frac{dh}{dt} = \frac{1}{4}\pi h^2 \frac{dh}{dt}$. Now we substitute in the known rates and variables: $-\frac{2\pi}{3} = \frac{1}{4}\pi(3)^2 \frac{dh}{dt} = -\frac{2\pi}{3} = \frac{9\pi}{4}\frac{dh}{dt} =$. . Solve for the dh/dt: $\frac{dh}{dt} = -\frac{2\pi}{3} \cdot \frac{4}{9\pi} = -\frac{8}{27}$. Therefore, the water level is changing at a rate of $-\frac{8}{27}$ cm/sec.

Common Formulas for Related Rates	
Circles	Triangles
$A = \pi r^2$	$A = \frac{1}{2}bh$
$C = 2\pi r$	$a^2 + b^2 = c^2$
Spheres	Cones
$V = \frac{4}{3}\pi r^3$	$V = \frac{1}{3}\pi r^2 h$
$SA = 4\pi r^2$	$SA = \pi r\sqrt{r^2 + h^2} + \pi r^2$
Cylinders	Rectangular Prism
$V = \pi r^2 h$	$V = lwh$
$SA = 2\pi rh + 2\pi r^2$	$SA = 2lw + 2lh + 2wh$

Practice 23

1) A 17 ft ladder is leaning against a wall and sliding towards the floor. The top of the ladder is sliding down the wall at a rate of 2 ft/sec. How fast is the base of the ladder sliding away from the wall when the base of the ladder is 8 ft from the wall?

2) A 13 ft ladder is leaning against a wall and sliding towards the floor. The foot of the ladder is sliding away from the base of the wall at a rate of 7 ft/sec. How fast is the top of the ladder sliding down the wall when the top of the ladder is 5 ft from the ground?

3) A 6 ft tall person is walking towards a 18 ft tall lamppost at a rate of 6 ft/sec. Assume the scenario can be modeled with right triangles. At what rate is the length of the person's shadow changing when the person is 13 ft from the lamppost?

4) A 5 ft tall person is walking away from a 17 ft tall lamppost at a rate of 5 ft/sec. Assume the scenario can be modeled with right triangles. At what rate is the length of the person's shadow changing when the person is 12 ft from the lamppost?

5) An observer stands 400 ft away from a launch pad to observe a rocket launch. The rocket blasts off and maintains a velocity of 600 ft/sec. Assume the scenario can be modeled as a right triangle. How fast is the observer to rocket distance changing when the rocket is 300 ft from the ground?

6) An observer stands 700 ft away from a launch pad to observe a rocket launch. The rocket blasts off and maintains a velocity of 400 ft/sec. Assume the scenario can be modeled as a right triangle. How fast is the angle of elevation (in radians/sec) from the observer to rocket changing when the rocket is 2400 ft from the ground?

7) Oil spilling from a ruptured tanker spreads in a circle on the surface of the ocean. The radius of the spill increases at a rate of 9 m/min. How fast is the area of the spill increasing when the radius is 6 m?

8) A crowd gathers around a movie star, forming a circle. The radius of the crowd increases at a rate of 5 ft/sec. How fast is the area taken up by the crowd increasing when the radius is 7 ft?

9) A perfect cube shaped ice cube melts so that the length of its sides are decreasing at a rate of 4 mm/sec. Assume that the block retains its cube shape as it melts. At what rate is the volume of the ice cube changing when the sides are 9 mm each?

10) A hypothetical cube grows so that the length of its sides are increasing at a rate of 4 m/min. How fast is the volume of the cube increasing when the sides are 8 m each?

11) A spherical balloon is inflated so that its radius increases at a rate of 3 cm/sec. How fast is the volume of the balloon increasing when the radius is 6 cm?

12) A spherical snowball melts so that its radius decreases at a rate of 3 in/sec. At what rate is the volume of the snowball changing when the radius is 7 in?

13) A conical paper cup is 10 cm tall with a radius of 10 cm. The bottom of the cup is punctured so that the water leaks out at a rate of $\frac{16\pi}{3}$ cm³/sec. At what rate is the water level changing when the water level is 7 cm?

14) A conical paper cup is 20 cm tall with a radius of 10 cm. The cup is being filled with water so that the water level rises at a rate of 4 cm/sec. At what rate is water being poured into the cup when the water level is 4 cm?

15) One diagonal of a rhombus is decreasing at a rate of 8 centimeters per minute and the other diagonal of the rhombus is increasing at a rate of 12 centimeters per minute. At a certain instant, the decreasing diagonal is 5 centimeters and the increasing diagonal is 7 centimeters. What is the rate of change of the area of the rhombus at that instant?

Note: $A = \frac{d_1 \cdot d_2}{2}$

16) The length of a rectangle is increasing at a rate of 6 cm/s and its width is increasing at a rate of 4 cm/s. When the length is 26 cm and the width is 12 cm, how fast is the area of the rectangle increasing?

17) One base of a trapezoid is decreasing at a rate of 8 kilometers per second and the height of the trapezoid is increasing at a rate of 5 kilometers per second. The other base of the trapezoid is fixed at 4 kilometers. At a certain instant, the decreasing base is 12 kilometers and the height is 2 kilometers. What is the rate of change of the area of the trapezoid at that instant?

$$A = \frac{1}{2}(b_1 + b_2)h$$

18) The side of the base of a square prism is increasing at a rate of 6 meters per second and the height of the prism is decreasing at a rate of 3 meters per second. At a certain instant, the base's side is 5 meters and the height is 8 meters. What is the rate of change of the volume of the prism at that instant?

$$V = s^2 \cdot h$$

19) A boat is being pulled into a dock by a rope attached to it and passing through a pulley on the dock positioned 6 meters higher than the boat. If the rope is being pulled in at a rate of 3 meters per second, how fast is the boat approaching the dock when it is 8 meters from the dock?

20) One leg of a right triangle is decreasing at a rate of 6 kilometers per hour and the other leg of the triangle is increasing at a rate of 4 kilometers per hour. At a certain instant, the decreasing leg is 8 kilometers and the increasing leg is 2 kilometers. What is the rate of change of the area of the right triangle at that instant?

21) The surface area of a cylinder is increasing at a rate of 9π square meters per minute. The height of the cylinder is fixed at 3 meters. At a certain instant, the surface area is 36π square meters. What is the rate of change of the volume of the cylinder at that instant?

22) The area of a square is increasing at a rate of 20 square meters per minute. At a certain instant, the area is 25 square meters. What is the rate of change of the perimeter of the square at that instant?

23) The volume of a cube is increasing at a rate of 24 cubic meters per second. At a certain instant, the volume is 27 cubic meters. What is the rate of change of the surface area of the cube at that instant?

24) The volume of a sphere is increasing at a rate of 24π cubic meters per hour. At a certain instant, the volume is 36π cubic meters. What is the rate of change of the surface area of the sphere at that instant?

Local Linearity

Local linearity describes a property of differentiable functions. When we zoom in on a function, the function looks linear in a small section around any point.

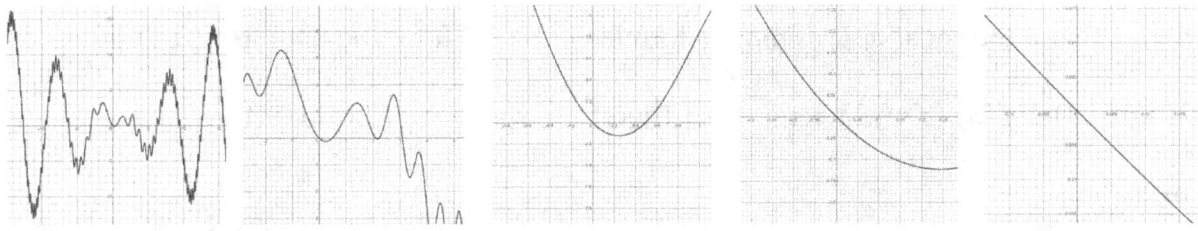

$f(x) = \sin x^2 + \sin x + x \sin x$ The function is zoomed in, further and further around the origin

This property allows us to approximate a differentiable function with a linear function around a specific point.

Recall the point-slope formula for a line: $y = m(x - x_1) + y_1$. When we apply this to our approximation method, we are trying to find a tangent line to a point on the function. The point will have coordinates of $(a, f(a))$. The slope of the tangent line at that point will be $f'(a)$. This can be used to derive the linear approximation for a function.

Linear Approximation: $L(x) = f'(a)(x - a) + f(a)$

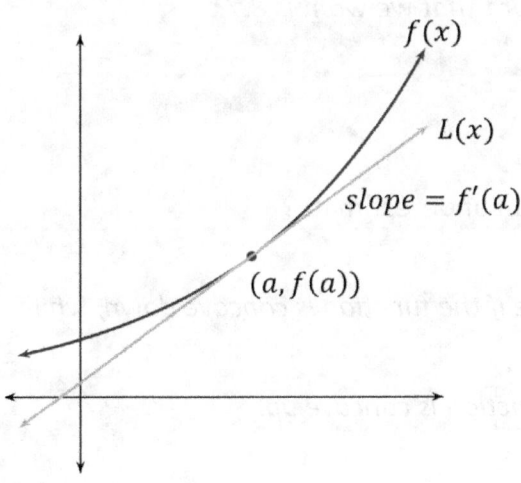

Ex. 1 Let f be a differentiable function with $f(3) = 6$ and $f'(3) = 4$.

What is the value of the approximation of $f(3.6)$ using the function's local linear approximation at $x = 3$.

Solution: We use the linear approximation formula using 3.6 for x and 3 for a. $L(3.4) = f'(3)(3.6 - 3) + f(3) = 4(.6) + 6 = 8.4$. Therefore, we have $f(3.6) \approx 8.4$ by linear approximation.

Ex. 2 The derivative of the function A is given by $A'(t) = 2 + t \sin t^2$, and $A(8.2) = 19.55$. If the linear approximation to is used to estimate $A(t)$ at what value of t does the linear approximation estimate that $A(t) = 16.6385$?

Solution: Use the linear approximation formula and substitute in what we know:
$f'(8.2)(t - 8.2) + f(8.2) = -5.823(t - 8.2) + 19.55 = 16.6385$ We need to solve this equation for t: $-5.823(t - 8.2) + 19.55 = 16.6385$ Distributing we get: $-5.823t + 47.7486 + 19.55 = 16.6385$ Simplifying we get: $-5.823t + 67.2986 = 16.6385 \Rightarrow -5.823t = -50.6601 \Rightarrow t = 8.7$ Therefore, the value for t that we want is 8.7.

Ex. 3 When will a linear approximation to a function be an underestimate?

Solution: *A linear approximation will be an overestimate if the function is concave down, while a linear approximation will be an underestimate if the function is concave up.*

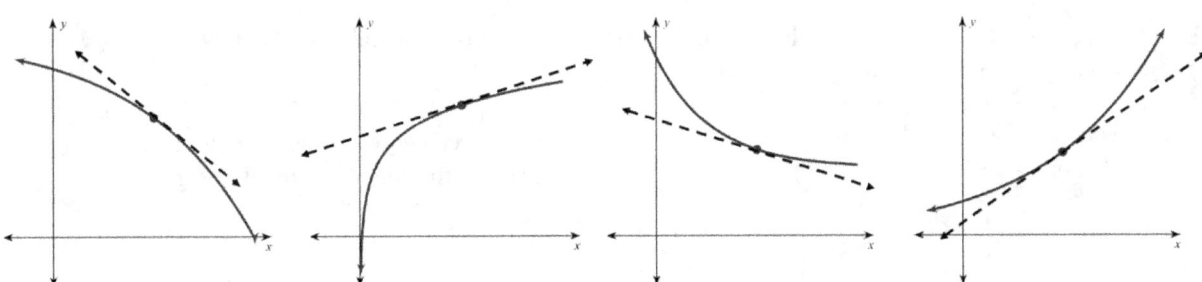

Linear Approximations for Concave Down Functions will **Overestimate**

Linear Approximations for Concave Down Functions will **Overestimate**

Practice 24

1) The local linear approximation to the function h at $x = 0$ is $y = 4x - 5$. What is the value of $h(0) + h'(0)$?

2) The local linear approximation to the function f at $x = 2$ is $y = 6x + 7$. What is the value of $f(2) + f'(2)$?

3) The local linear approximation to the function f at $x = 3$ is $y = 3x - 8$. What is the value of $f(3) + f'(3)$?

4) The local linear approximation to the function f at $x = 4$ is $y = 5x - 8$. What is the value of $f(4) + f'(4)$?

5) Let g be a differentiable function with $g(2) = 4$ and $g'(2) = 9$
What is the value of the approximation of $g(2.4)$ using the function's local linear approximation at $x = 2$

6) Let f be a differentiable function with $f(-3) = 0$ and $f'(-3) = 7$
What is the value of the approximation of $f(-3.2)$ using the function's local linear approximation at $x = -3$

7) Let f be a differentiable function with $f(2) = 4$ and $f'(2) = 10$
What is the value of the approximation of $f(2.4)$ using the function's local linear approximation at $x = 2$

8) Let f be a differentiable function with $f(3) = 6$ and $f'(3) = \dfrac{1}{2}$
What is the value of the approximation of $f(3.2)$ using the function's local linear approximation at $x = 3$

9) The derivative of the function A is given by $A'(t) = 3 + 10e^{0.2\sin t}$, and $A(1.4) = 8.5$. If the linear approximation to $A(t)$ at $t = 1.4$ is used to estimate $A(t)$, at what value of t does the linear approximation estimate that $A(t) = 15.71$?

A) 1.587 B) 1.875
C) 2.732 D) 2.435

10) The derivative of the function A is given by $A'(t) = te^{0.1t^2 + 0.4}$, and $A(1.8) = 10.3$. If the linear approximation to $A(t)$ at $t = 1.8$ is used to estimate $A(t)$, at what value of t does the linear approximation estimate that $A(t) = 12.16$?

A) 3.7 B) 3.2
C) 48.8 D) 2.3

11) The derivative of the function A is given by $A'(t) = \tan^{-1} t$, and $A(2.4) = 1.9$. If the linear approximation to $A(t)$ at $t = 2.4$ is used to estimate $A(t)$, at what value of t does the linear approximation estimate that $A(t) = 2.78$?

A) 3.35 B) 3.15
C) 2.15 D) 2.35

12) The derivative of the function A is given by $A'(t) = \dfrac{t^3 + 1}{t - 1}$, and $A(3.2) = 20.82$. If the linear approximation to $A(t)$ at $t = 3.2$ is used to estimate $A(t)$, at what value of t does the linear approximation estimate that $A(t) = 53.54$?

A) 5.7 B) 5.5
C) 5.3 D) 5.1

13. The locally linear approximation of the differentiable function f at $x = 3$ is used to approximate the value of $f(2.6)$. The approximation at $x = 2.6$ is an overestimate of the corresponding function value at $x = 2.6$. Which of the following could be the graph of f?

    (The point at $x = 3$ is marked)

    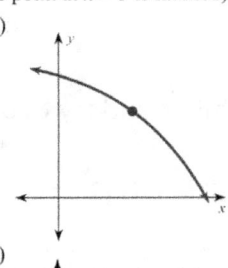

14. The locally linear approximation of the differentiable function f at $x = 3$ is used to approximate the value of $f(3.4)$. The approximation at $x = 3.4$ is an overestimate of the corresponding function value at $x = 3.4$. Which of the following could be the graph of f?

    (The point at $x = 3$ is marked)

    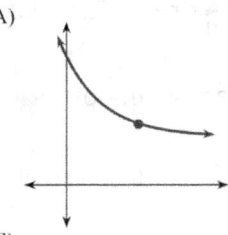
    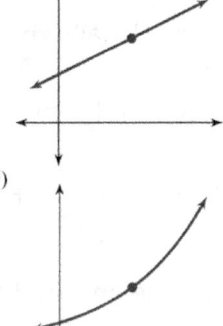

15. The locally linear approximation of the differentiable function f at $x = 3$ is used to approximate the value of $f(2.6)$. The approximation at $x = 2.6$ is an underestimate of the corresponding function value at $x = 2.6$. Which of the following could be the graph of f?

    (The point at $x = 3$ is marked)

    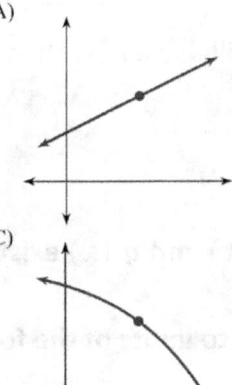
    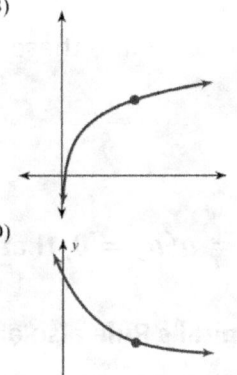

16. The locally linear approximation of the differentiable function f at $x = 3$ is used to approximate the value of $f(3.4)$. The approximation at $x = 3.4$ is an underestimate of the corresponding function value at $x = 3.4$. Which of the following could be the graph of f?

    (The point at $x = 3$ is marked)

    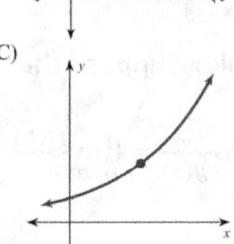
    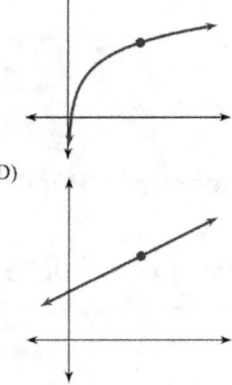

Indeterminate Forms and L'Hôpital's Rule

An indeterminate form is an algebraic expression that when we try to evaluate the limit, we get a form involving infinity or zero that can not be determined directly. The most common indeterminate forms are $\frac{0}{0}$ or $\frac{\infty}{\infty}$. It should be easy to see why $\frac{0}{0}$ is considered an indeterminate form. Usually zero divided by any other number is zero. If we take a number and try to divide by zero, it is undefined. Other indeterminate forms include the following: $0 \cdot \infty$, $\infty - \infty$, 0^0, 1^∞, and ∞^0. It should be noted that the form $\frac{1}{0}$ is not considered an indeterminate form, because any limit of $\frac{f}{g}$ of the form $\frac{1}{0}$ does one of three things: 1) goes to infinity 2) goes to negative infinity or 3) does not exist.

We have already used algebraic manipulation to simplify limits of indeterminate form such as: $\lim_{x \to 2} \frac{x^2+2x-8}{x-2}$. If we try direct substitution would yield the form: $\frac{0}{0}$. However, we can factor and simplify to evaluate: $\lim_{x \to 2} \frac{x^2+2x-8}{x-2} = \lim_{x \to 2} \frac{(x+4)(x-2)}{x-2} = \lim_{x \to 2}(x+4) = 6$.

There are some indeterminate forms of limits which cannot be simplified by algebraic manipulation or any other previous technique. To evaluate these, we will use a new theorem.

Theorem: L'Hôpital's Rule: Suppose that $f(a) = g(a) = 0$, that $f'(x)$ and $g'(x)$ exist, and that $g'(a) \neq 0$. Then $\lim_{x \to a} \frac{f(x)}{g(x)} = \lim_{x \to a} \frac{f'(x)}{g'(x)}$. L'Hôpital's Rule also applies to limits of the form $\frac{\infty}{\infty}$

Note that these indeterminate forms, are not real numbers and so you should never write that a limit *equals* any of these forms. You state, "the limit is of the form… and we apply L'Hôpital's Rule."

Ex. 1 Evaluate $\lim\limits_{x \to 0} \frac{3(e^x - e^{-x})}{x}$

Solution: First try direct substitution. $\lim\limits_{x \to 0} \frac{3(e^x - e^{-x})}{x}$ is of the form $\frac{0}{0}$. We apply L'Hôpital's Rule:

$\lim\limits_{x \to 0} \frac{3(e^x - e^{-x})}{x} = \lim\limits_{x \to 0} \frac{3(e^{-x} + e^{-x})}{1} = 6$. *Therefore, the limit is 6.*

Ex. 2 Evaluate $\lim\limits_{x \to \infty} \frac{\ln x^4}{\ln(x+5)^3}$

Solution: First try direct substitution. $\lim\limits_{x \to \infty} \frac{\ln x^4}{\ln(x+5)^3}$ is of the form $\frac{\infty}{\infty}$. We apply L'Hôpital's Rule:

$\lim\limits_{x \to \infty} \frac{\ln x^4}{\ln(x+5)^3} = \lim\limits_{x \to \infty} \frac{4/x}{3/(x+5)} = \lim\limits_{x \to \infty} \frac{4(x+5)}{3x}$. This is still in an indeterminate form, so we can apply

L'Hôpital's Rule again: $\lim\limits_{x \to \infty} \frac{4(x+5)}{3x} = \lim\limits_{x \to \infty} \frac{4}{3} = \frac{4}{3}$. *Therefore, the limit is $\frac{4}{3}$.*

Ex. 3 Evaluate $\lim\limits_{x \to 0} \frac{2(e^x - x)}{x^2}$

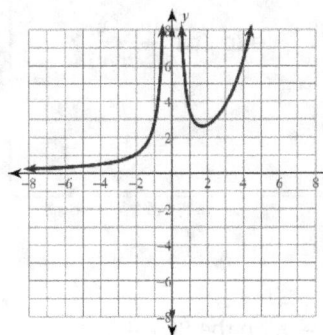

Solution: First try direct substitution. $\lim\limits_{x \to 0} \frac{2(e^x - x)}{x^2}$ is of the form $\frac{2}{0}$.

This is undefined, but **not an indeterminate form**, so we **cannot** apply L'Hôpital's Rule. We must use other techniques, such as graphing, or using a calculator's table to evaluate. Usually when you have the form $\frac{a}{0}$, where $a \neq 0$, the function has a vertical asymptote at the value we are approaching. Here, the graph shows the vertical asymptote and we can see the left and right side limits both approach positive infinity. Therefore, $\lim_{x \to 0} \frac{2(e^x - x)}{x^2} = \infty$.

Other indeterminate forms[12] may be evaluated using L'Hôpital's Rule by making an appropriate algebraic substitution first. Below is a table showing the transformations useful for these additional indeterminate forms.

Indeterminate Form	Conditions	Transformation to $\frac{0}{0}$	Transformation to $\frac{\infty}{\infty}$
$0 \cdot \infty$	$\lim_{x \to a} f(x) = 0$, $\lim_{x \to a} g(x) = \infty$ Find $\lim_{x \to a} f(x)g(x)$	$\lim_{x \to a} \frac{f(x)}{1/g(x)}$	$\lim_{x \to a} \frac{g(x)}{1/f(x)}$
$\infty - \infty$	$\lim_{x \to a} f(x) = \infty$, $\lim_{x \to a} g(x) = \infty$ Find $\lim_{x \to a}[f(x) - g(x)]$	$\lim_{x \to a} \frac{1/g(x) - 1/f(x)}{1/(f(x)g(x))}$	$\ln \lim_{x \to a} \frac{e^{f(x)}}{e^{g(x)}}$
0^0	$\lim_{x \to a} f(x) = 0^+$, $\lim_{x \to a} g(x) = 0$ Find $\lim_{x \to a} f(x)^{g(x)}$	$\exp \lim_{x \to a} \frac{g(x)}{1/\ln f(x)}$	$\exp \lim_{x \to a} \frac{\ln f(x)}{1/g(x)}$

---

[12] These additional forms are not tested on the AP Calculus AB exam, but are seen on the BC exam

	$\lim_{x\to a} f(x) = 1$,		
1^∞	$\lim_{x\to a} g(x) = \infty$	$\exp \lim_{x\to a} \dfrac{\ln f(x)}{1/g(x)}$	$\exp \lim_{x\to a} \dfrac{g(x)}{1/\ln f(x)}$
	Find $\lim_{x\to a} f(x)^{g(x)}$		
	$\lim_{x\to a} f(x) = \infty$,		
∞^0	$\lim_{x\to a} g(x) = 0$	$\exp \lim_{x\to a} \dfrac{g(x)}{1/\ln f(x)}$	$\exp \lim_{x\to a} \dfrac{\ln f(x)}{1/g(x)}$
	Find $\lim_{x\to a} f(x)^{g(x)}$		

Practice 25

1) $\lim_{x\to 0} \dfrac{e^{2x}-1}{5x}$

2) $\lim_{x\to 1} \dfrac{\ln x^2}{x^2-1}$

3) $\lim_{r\to\infty} \dfrac{e^{2r}}{r}$

4) $\lim_{s\to\infty} \dfrac{\ln s}{2s}$

5) $\lim_{x\to 0} \dfrac{x^2}{e^x-1-x}$

6) $\lim_{x\to 0} \dfrac{4(e^x-1-x)}{1-\cos x}$

7) $\lim\limits_{x \to \infty} \dfrac{x}{\ln x}$

8) $\lim\limits_{x \to 0} \dfrac{e^{4x} - 1}{2x}$

9) $\lim\limits_{x \to 1} \dfrac{\ln x}{x - 1}$

10) $\lim\limits_{x \to 0^+} \dfrac{2\ln \tan x}{\ln \sin x}$

11) $\lim\limits_{x \to 1} \dfrac{4 \ln x^2}{x^2 - 1}$

12) $\lim\limits_{x \to \infty} \dfrac{x}{e^{2x}}$

13) $\lim\limits_{x \to -3} -\dfrac{x + 3}{x^2 + 7x + 12}$

14) $\lim\limits_{x \to 0} \dfrac{3(e^x - e^{-x})}{\sin(2x)}$

15) $\lim\limits_{x \to \infty} \dfrac{\ln(x + 3)^4}{\ln x^2}$

16) $\lim\limits_{x \to \infty} \dfrac{\ln x^5}{\ln(x + 1)^4}$

17) $\lim\limits_{x \to 0} \dfrac{e^x - 1 - x}{1 - \cos x}$

18) $\lim\limits_{x \to 0} \dfrac{4(e^x - 1 - x)}{1 - \cos x}$

19) $\lim\limits_{x \to \infty} \dfrac{e^x}{2x^2}$

20) $\lim\limits_{x \to \infty} \dfrac{\ln (x+1)^2}{\ln x^2}$

21) $\lim\limits_{x \to \infty} \dfrac{\ln (x+5)^4}{\ln x^4}$

22) $\lim\limits_{x \to \infty} \dfrac{\ln (x+4)^3}{\ln x^4}$

Determine if L'Hôpital's Rule can be applied. If it can be applied write YES, if it cannot be applied, write NO. Determine the limit.

23) $\lim\limits_{x \to 1^+} \dfrac{2x}{\ln x}$

24) $\lim\limits_{x \to 0^+} \dfrac{2x}{\ln x}$

25) $\lim\limits_{x \to 0^+} \dfrac{3 \ln \tan x}{\ln \sin x}$

26) $\lim\limits_{x \to 0^+} \dfrac{4 \ln \tan x}{\ln \cos x}$

27) $\lim\limits_{x \to 1} \dfrac{3(x-1)}{\ln x}$

28) $\lim\limits_{x \to 0^+} \dfrac{2(e^x + e^{-x})}{x}$

Advanced: Use a transformation and L'Hôpital's Rule to evaluate:

29) $\lim\limits_{x \to \infty} 5x\sin\dfrac{1}{x}$

30) $\lim\limits_{x \to 0^+} 2x\ln\dfrac{1}{x}$

31) $\lim\limits_{x \to \frac{\pi}{2}} (\sec x - \tan x)$

32) $\lim\limits_{x \to \infty} \left(\dfrac{x^2}{x-1} - \dfrac{x^2}{x+1}\right)$

33) $\lim\limits_{x \to \infty} x^{\frac{1}{x}}$

34) $\lim\limits_{x \to 0} (2x+1)^{\frac{1}{x}}$

35) $\lim\limits_{x \to 0} (x+1)^{\frac{1}{x}}$

36) $\lim\limits_{x \to 0^+} 5x^{2x}$

Applying Derivatives to Analyze Functions

In this next section, we will examine how calculus can be used to analyze functions.

Intermediate Value Theorem

Intermediate Value Theorem: If f is a continuous function on the interval $[a, b]$, and L is any number between $f(a)$ and $f(b)$, where $f(a) \neq f(b)$, then there is a number c such that $f(c) = L$.

For this theorem to work, the function must be continuous, which means you can draw it without lifting your pencil. Since differentiable functions are also continuous, this theorem can also be applied to derivatives. It is called the *intermediate* value theorem, referring to *between* two values.

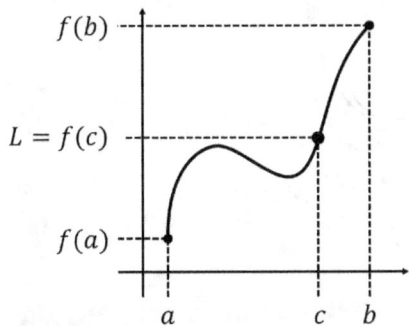

Ex. 1 Use the table below to answer the following questions:

x	−6	−4	0	2	4	6	10
$f(x)$	−6	−2	4	−4	8	12	−2

1. Does there exist some x in the interval $[-6, -4]$ such that $f(x) = -3$?

2. Does there exist some x in the interval $[4,6]$ such that $f(x) = 10$?

3. Does there exist some x in the interval $[-4,0]$ such that $f(x) = 2$?

4. What is the minimum number of zeros $f(x)$ has?

Page |161

Solutions:

1. Does there exist some x in the interval $[-6, -4]$ such that $f(x) = -3$?

   Yes, because -3 is between the y-values of -6 and -2.

2. Does there exist some x in the interval $[4,6]$ such that $f(x) = 6$?

   No, because 6 is not between the y-values of 8 and 12.

3. Does there exist some x in the interval $[-4,0]$ such that $f(x) = 2$?

   Yes, because -2 is between the y-values of -2 and 4.

4. What is the minimum number of zeros $f(x)$ has?

   The minimum number of zeros is 4, because we count the sign changes in the y-values.

   Specifically there are zeros in the intervals: (-4, 0), (0, 2), (2, 4), and (6, 10)

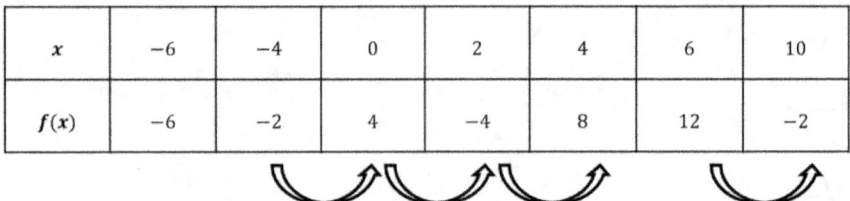

Sign changes tells us there is a zero in the interval

Ex. 2 Use the table below to answer the following questions:

x	-6	-4	0	2	4	6	10
$f'(x)$	6	-2	-4	4	8	-12	-2

1. Does there exist some x in the interval $[-6, -4]$ such that $f'(x) = -3$?

2. Does there exist some x in the interval $[4,6]$ such that $f'(x) = -10$?

3. Does there exist some x in the interval $[-4,0]$ such that $f'(x) = -3$?

4. What is the minimum number of critical points of $f(x)$?

Solutions:

1. Does there exist some x in the interval $[-6, -4]$ such that $f'(x) = -3$?

   No, because -3 is not between 6 and -2

2. Does there exist some x in the interval $[4,6]$ such that $f'(x) = -10$?

   Yes, because -10 is between 8 and -12

3. Does there exist some x in the interval $[-4,0]$ such that $f'(x) = -3$?

   Yes, because -3 is between -2 and -4

4. What is the minimum number of critical points of $f(x)$?

   The minimum number of critical points, or zeros, are the number of sign changes which is 3 sign changes at (-6, -4), (0, 2) and (4,6).

Practice 26

Use the table below to answer the following questions:

x	-6	-4	0	2	4	6	10
$f(x)$	-4	-8	4	6	-8	-2	4

1. Does there exist some x in the interval $[-6, -4]$ such that $f(x) = -2$?

2. Does there exist some x in the interval $[4,6]$ such that $f(x) = -6$?

3. Does there exist some x in the interval $[-4,0]$ such that $f(x) = 2$?

4. Does there exist some x in the interval $[0,2]$ such that $f(x) = 2$?

5. Does there exist some x in the interval $[6,10]$ such that $f(x) = 8$

6. What is the minimum number of zeros $f(x)$ has? Where are there possible zeros?

Mean Value Theorem

The Mean Value Theorem basically states that on a differentiable function, there exists at least one point where the instantaneous rate of change will be the same as the average rate of change. One application, is that if you travel 320 miles in 4 hours, then there must have been at least one time that you were traveling exactly 80 mph.

Mean Value Theorem: Let f be a function be continuous on the closed interval $[a, b]$ and differentiable on the open interval (a, b), then there is a number c in (a, b) such that $f'(c) = \frac{f(b)-f(a)}{b-a}.$

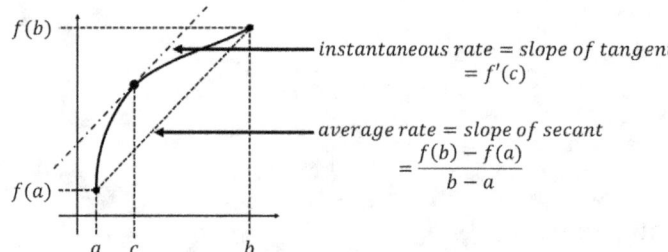

This theorem guarantees that if a function is differentiable then there exists at least one tangent line parallel to the secant line. There could possibly be more than one. Of course, if the function is not smooth, differentiable then we are not guaranteed such a tangent line exists as seen in the picture below.

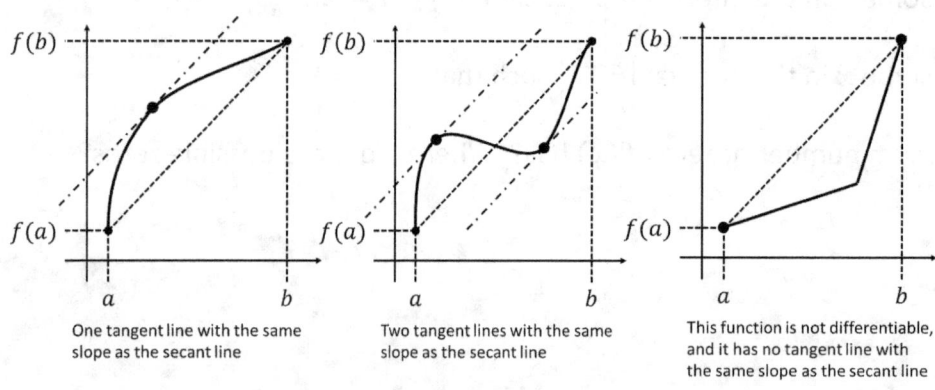

Ex. 1 Let $f(x) = x^2 + 6x + 11$ Find c that satisfies the Mean Value Theorem for f on the interval $[-5, -2]$.

Solution: For the Mean Value Theorem to work we first should check the conditions for the theorem. We must make sure it is continuous on the closed interval $[-5, -2]$ and differentiable on the open interval $(-5, -2)$. For this example, we have a polynomial. Polynomials are continuous and differentiable everywhere, so the conditions are satisfied.

Next, we find the average rate of change:

$$\frac{f(b)-f(a)}{b-a} = \frac{f(-2)-f(-5)}{-2--5} = \frac{f(-2)-f(-5)}{-2--5} = \frac{3-6}{-2--5} = \frac{-3}{3} = -1$$

So, the average rate of change over the interval is -1. Now we take the derivative of our function. $f'(x) = 2x + 6$. Set this equal to the rate of change: $2x + 6 = -1$. Solving for x gives: $x = -\frac{7}{2}$. This means that the instantaneous rate of change at $x = -\frac{7}{2}$ is the same as the average rate of change. If we wanted to, we could find the y-coordinate as well by evaluating the function at x = -7/2 to get the point $(-\frac{7}{2}, \frac{9}{4})$.

Ex. 2 Determine if the Mean Value Theorem can be applied. If it can, find all values of c that satisfy the theorem. If it cannot, explain why not. Let $f(x) = -(6x + 36)^{\frac{2}{3}}$ on the interval $[-7, -3]$.

Solution: For the Mean Value Theorem to work we first should check the conditions for the theorem. We must make sure it is continuous on the closed interval $[-7, -3]$ and differentiable on the open interval $(-7, -3)$. For this example, we have a function with a rational exponent, so we need to check. Graphing the function is the easiest way to check:

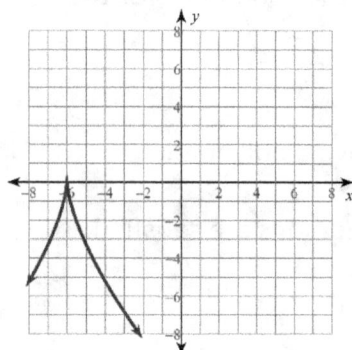

This function is not differentiable at -6 because it has a cusp at that point. Therefore, the mean value theorem does not apply

Practice 27

For each problem, find the values of c that satisfy the Mean Value Theorem.

1) Let $f(x) = -x^2 + 2x + 4$
   Find c that satisfies the Mean Value Theorem for f on the interval $[2, 4]$

2) Let $f(x) = 2x^2 + 4x + 1$
   Find c that satisfies the Mean Value Theorem for f on the interval $[-3, 1]$

3) Let $f(x) = -x^2 - 8x - 15$
   Find c that satisfies the Mean Value Theorem for f on the interval $[-6, -2]$

4) Let $f(x) = \dfrac{x^2}{2} + x + \dfrac{3}{2}$
   Find c that satisfies the Mean Value Theorem for f on the interval $[-3, -1]$

5) $y = \dfrac{x^2}{2} + 2$; $[-3, 2]$

6) $y = x^2 + 6x + 6$; $[-4, -1]$

7) $y = -x^3 + 4x^2 - 7$; $[0, 2]$

8) $y = -x^3 + 7x^2 - 11x - 1$; $[1, 3]$

9) $y = -x^3 + 3x^2 - 4$; $[-1, 3]$

10) $y = -x^3 + 4x^2 - 2$; $[-1, 3]$

11) $y = \dfrac{x^2}{4x + 4}$; $[-5, -2]$

12) $y = \dfrac{x^2 - 9}{2x}$; $[1, 4]$

13) $y = -\dfrac{x^2}{3x + 6}$; $[-1, 3]$

14) $y = -\dfrac{x^2}{4x - 4}$; $[-4, 0]$

15) $y = -(-3x + 12)^{\frac{1}{2}}$; [0, 4]

16) $y = (3x + 15)^{\frac{1}{2}}$; [−5, 0]

For each problem, determine if the Mean Value Theorem can be applied. If it can, find all values of c that satisfy the theorem. If it cannot, explain why not. You may use the provided graph to sketch the function.

17) $y = \dfrac{x^2}{4x - 8}$; [1, 3]

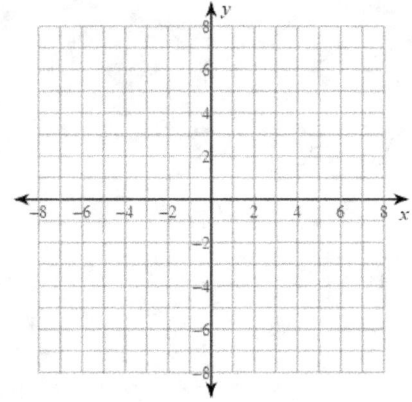

18) $y = -(2x - 8)^{\frac{2}{3}}$; [1, 4]

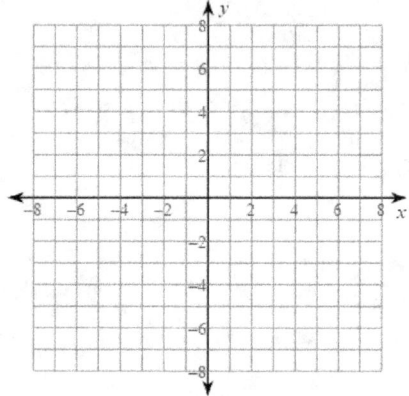

19) $y = \dfrac{x^2 - 4}{3x}$; [−3, −1]

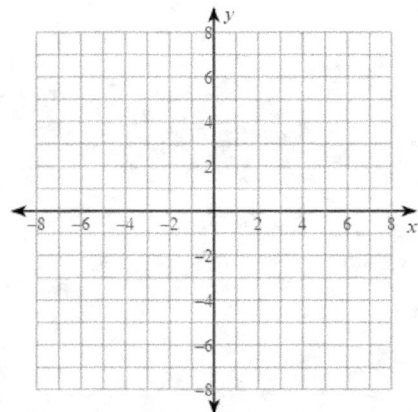

20) $y = -(2x - 8)^{\frac{2}{3}}$; [3, 6]

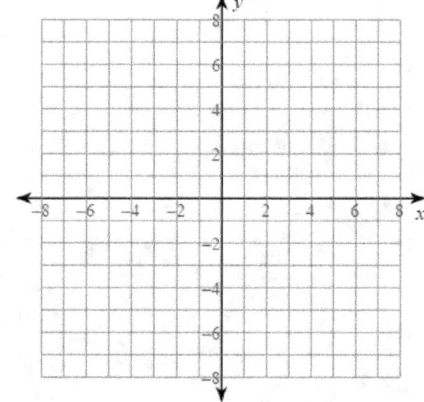

Extrema

Extrema is the plural of extreme, which is any maximum (high point) or minimum (low point). Extrema may be relative (also called local) or absolute (also called global). There may be several relative extrema. There can be at most one global maximum and at most one global minimum.

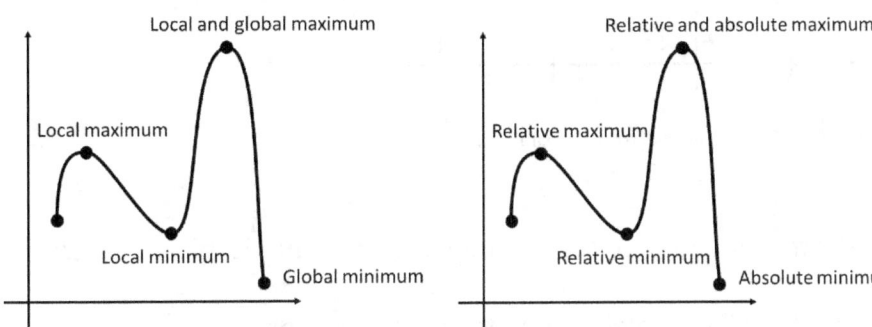

Extreme Value Theorem

The Extreme Value Theorem basically states that on a continuous function on a closed interval, there exists an absolute maximum and absolute minimum, or a highest point and lowest point.

Extreme Value Theorem: Let f be a function be continuous on the closed interval $[a, b]$ (a, b), then f contains an absolute maximum value $f(c)$ and an absolute minimum value $f(d)$ at numbers c and d in $[a, b]$.

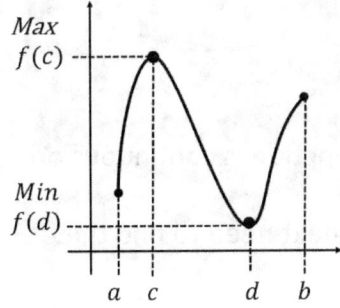

It is important that the interval be **closed** and **continuous** to guarantee that both the absolute maximum and absolute minimum exist. See below for examples.

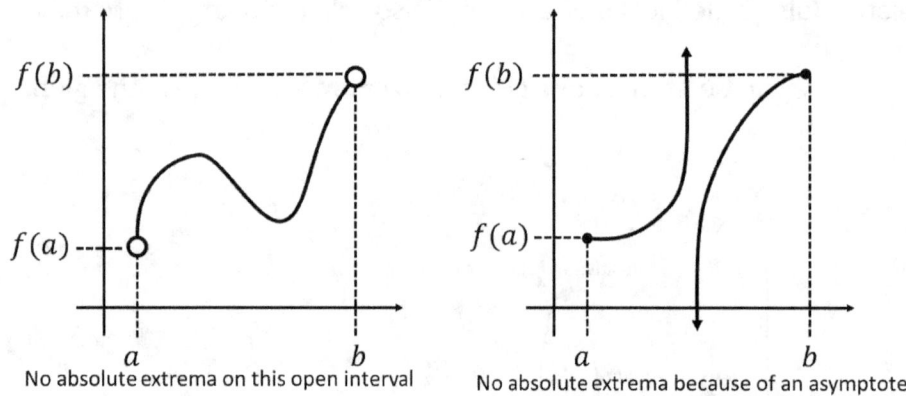

No absolute extrema on this open interval

No absolute extrema because of an asymptote

When trying to find absolute extrema, we need to test all critical points, that is, places where the derivative is either zero or fails to exist. We also need to check the endpoints.

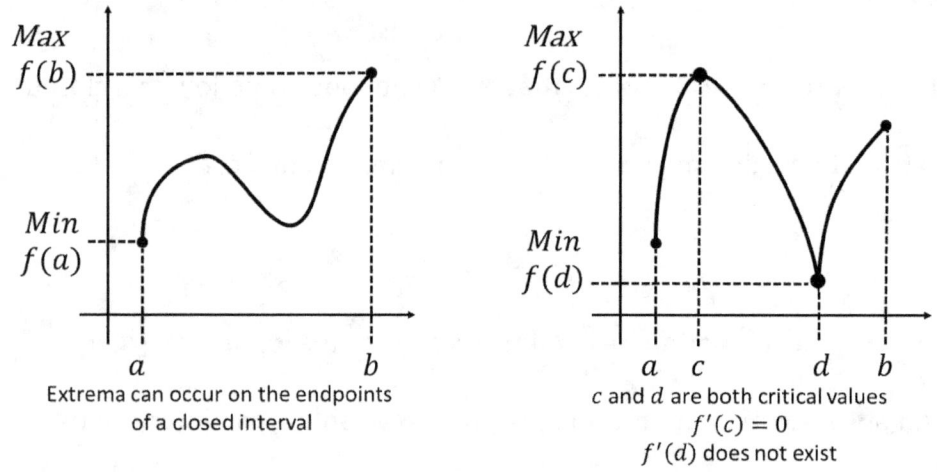

Extrema can occur on the endpoints of a closed interval

c and d are both critical values
$f'(c) = 0$
$f'(d)$ does not exist

<u>Ex. 1</u> Let $f(x) = -x^3 + 4x^2 - 4$ Find all points of absolute minima and maxima on the interval $[-1, 4]$.

<u>Solution:</u> First we check that we have a closed interval and that the function is continuous on the closed interval. Polynomial functions are always continuous. We next need to find the

critical values by finding the derivative: $f'(x) = -3x^2 + 8x$. The derivative is always defined, so we need to set it to zero: $-3x^2 + 8x = 0$. Solving: $x(-3x + 8) = 0$, so $x = 0$ or $x = \frac{8}{3}$.

Next, we evaluate the function at each endpoint and critical value:

$f(-1) = -(-1)^3 + 4(-1)^2 - 4 = 1$

$f(0) = -(0)^3 + 4(0)^2 - 4 = -4$ Minimum

$f\left(\frac{8}{3}\right) = -\left(\frac{8}{3}\right)^3 + 4\left(\frac{8}{3}\right)^2 - 4 = \frac{148}{27}$ Maximum

$f(4) = -(4)^3 + 4(4)^2 - 4 = -4$ Minimum

Therefore, the function has a minimum at the points $(0, -4)$ and $(4, -4)$, while it has a maximum at $\left(\frac{8}{3}, \frac{148}{27}\right)$.

Practice 28

For each problem, find all points of absolute minima and maxima on the given interval.

1) $y = -\frac{x^2}{2} + 2x + 1;\ [1, 6]$

2) $y = -\frac{x^2}{2} - 4x - 7;\ [-7, -5]$

3) $y = -2x^2 - 4x - 4;\ [-2, 0]$

4) $y = -\frac{x^2}{2} - 4x - 3;\ [-7, -4]$

5) $y = -x^3 + 2x^2 + 3$; $[0, 2]$

6) $y = -x^3 + 3x^2$; $[0, 3]$

7) $y = -x^3 + 4x^2 - 2$; $[-1, 1]$

8) $y = -x^3 + x^2 + 1$; $[-1, 1]$

9) $y = -x^4 + 4x^2 - 2$; $[-1, 1]$

10) $y = -x^4 + x^2 - 4$; $[-1, 1]$

11) $y = x^4 - 3x^2$; $[0, 2]$

12) $y = -x^4 + 4x^2 - 5$; $[-2, 1]$

13) $y = \dfrac{1}{x^2 + 1}$; $[-4, -2]$

14) $y = \dfrac{2}{x^2 - 16}$; $[-2, 2]$

15) $y = \dfrac{16x}{x^2 + 16}$; [2, 6]

16) $y = -\dfrac{20}{x^2 + 4}$; [−2, 2]

17) $y = (6x + 36)^{\frac{1}{3}}$; [−5, 1]

18) $y = (6x + 30)^{\frac{1}{2}}$; [−3, 3]

19) $y = \dfrac{1}{4}(x + 2)^{\frac{8}{3}} - 4(x + 2)^{\frac{2}{3}} - 2$; [−2, 3]

20) $y = -\dfrac{3}{16}(x + 2)^{\frac{4}{3}} + \dfrac{3}{2}(x + 2)^{\frac{1}{3}} + 2$; [0, 4]

21) $y = -2\csc(x)$; $[\dfrac{\pi}{6}, \dfrac{\pi}{4}]$

22) $y = -\cot(x)$; $[-\dfrac{\pi}{6}, \dfrac{3\pi}{4}]$

23) $y = -2\sin(2x)$; $[-\frac{\pi}{4}, 0]$

24) $y = 2\cos(x)$; $[\frac{\pi}{3}, \frac{\pi}{2}]$

25) $y = \cos(x)$; $[-\frac{\pi}{2}, -\frac{\pi}{3}]$

26) $y = -2\sin(x)$; $[-\frac{3\pi}{4}, 0]$

Intervals of Increasing and Decreasing

We can use the derivative to determine intervals where a function is increasing (going uphill as you go left to right) and decreasing (going downhill left to right). When a function switches from increasing to decreasing or vice versa, there will be a critical point. Remember a critical point is where the derivative is either zero or doesn't exists.

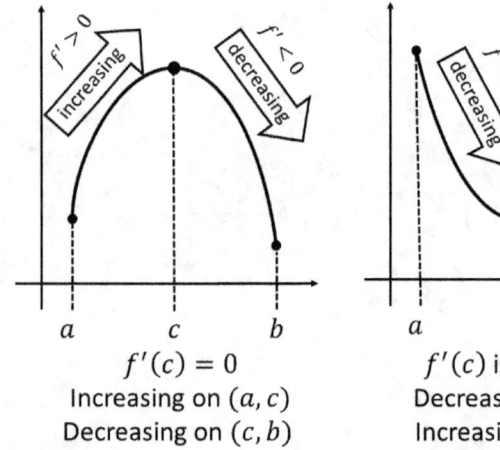

Steps to find intervals of increasing and decreasing: Find all x values where the derivative is either zero or undefined. Evaluate the derivative at a value in each interval.

Ex. 1 Let $f(x) = -2x^2 + 12x - 14$ Find all critical values and discontinuities. Determine intervals of increasing and intervals of decreasing.

Solution: First, note that since we have a polynomial there are no discontinuities. Now we take the derivative: $f'(x) = -4x + 12$. Set to zero and solve for x: $-4x + 12 = 0$ gives $x = 3$.

We test the interval by taking the derivative at a value before 3 and one after 3. Evaluating at 0 is usually easy to do. $f'(0) = -4(0) + 12 = 12$. The derivative at 0 is positive. $f'(10) = -4(10) + 12 = -28$. The derivative at 10 is negative. We can organize this on a number line like below. We can also graph the function to confirm.

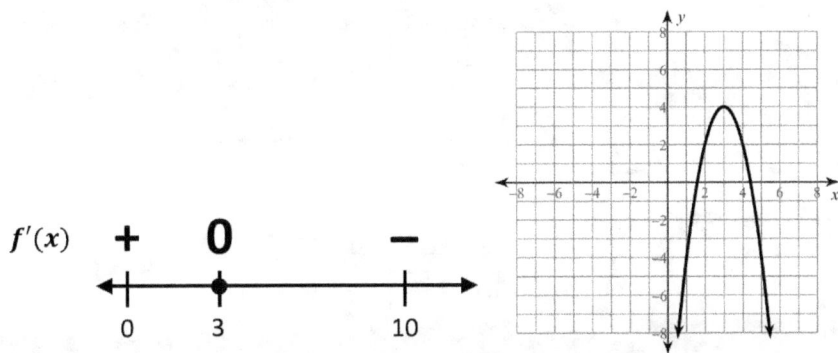

Therefore, the intervals of increasing are $(-\infty, 3)$ and intervals of decreasing are $(3, \infty)$.

Ex. 2 Let $f(x) = \frac{2}{x^2-16}$ Find all critical values and discontinuities. Determine intervals of increasing and intervals of decreasing.

Solution: First, note that since we have a rational function so we need to check for discontinuities. The function is undefined when the denominator is 0. This happens when x = -4 and x = 4. Now we take the derivative: $f(x) = \frac{2}{x^2-16} = 2(x^2-16)^{-1}$. $f'(x) = -2(x^2-16)^{-2}(2x) = -\frac{4x}{(x^2-16)^2}$. Set to zero, a rational function is equal to zero when the numerator is zero. Set the numerator to zero and solve for x: $-4x = 0$ gives $x = 0$.

We have three critical values to check between $-4, 0,$ and 4. $f'(-10) = -\frac{4(-10)}{((-10)^2-16)^2} = \frac{5}{882}$. $f'(-2) = -\frac{4(-2)}{((-2)^2-16)^2} = \frac{1}{18}$. $f'(2) = -\frac{4(2)}{((2)^2-16)^2} = -\frac{1}{18}$. $f'(10) = -\frac{4(10)}{((10)^2-16)^2} = -\frac{5}{882}$. We can again organize this on a number line. I use asterisks to note where asymptotes/discontinuities are located. We can also graph the function to confirm.

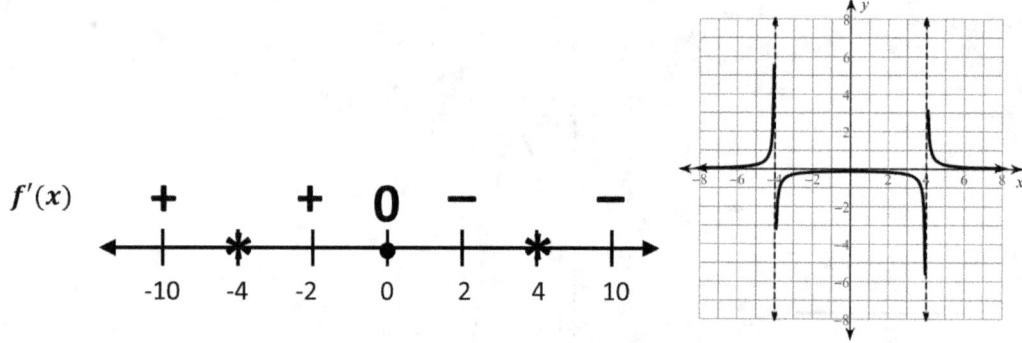

Therefore, the intervals of increasing are $(-\infty, -4) \cup (-4, 0)$ and intervals of decreasing are $(0,4) \cup (4, \infty)$.

Practice 29

For each problem, find the x-coordinates of all critical points, find all discontinuities, and find the open intervals where the function is increasing and decreasing.

1) $f(x) = 2x^2 + 4x + 2$

2) $f(x) = x^2 + 6x + 4$

3) $f(x) = x^2 - 4x + 6$

4) $f(x) = -x^2 + 4x - 4$

5) $f(x) = x^3 - 3x^2 + 5$

6) $f(x) = -x^3 + 14x^2 - 64x + 95$

7) $f(x) = -x^3 + 2x^2 + 3$

8) $f(x) = -x^3 + x^2$

9) $f(x) = x^4 - 3x^2 + 3$

10) $f(x) = x^4 - 2x^2 - 3$

11) $f(x) = x^4 - x^2 - 3$

12) $f(x) = -x^4 + 3x^2 + 1$

13) $f(x) = -x^5 + 2x^3 + 1$

14) $f(x) = x^5 - 3x^3 - 1$

15) $f(x) = -x^5 + 3x^3$

16) $f(x) = -x^5 + 3x^3 - 2$

17) $f(x) = -2\sin(x)$; $[-\pi, \pi]$

18) $f(x) = 2\cot(2x)$; $[-\pi, \pi]$

19) $f(x) = -2\cos(x)$; $[-\pi, \pi]$

20) $f(x) = -2\tan(x)$; $[-\pi, \pi]$

21) $f(x) = -\dfrac{3x}{x-3}$

22) $f(x) = -\dfrac{2}{x+3}$

23) $f(x) = -\dfrac{1}{x+3}$

24) $f(x) = \dfrac{x^2}{4x-8}$

25) $f(x) = -\dfrac{x}{x^2 - 1}$

26) $f(x) = -\dfrac{4x^2 - 4}{x^3}$

27) $f(x) = -\dfrac{3}{x^2 - 16}$

28) $f(x) = -\dfrac{16x}{x^2 + 16}$

29) $f(x) = -(2x - 10)^{\frac{1}{3}}$

30) $f(x) = (2x + 12)^{\frac{2}{3}}$

31) $f(x) = -\dfrac{3}{16}(x - 2)^{\frac{4}{3}} - \dfrac{3}{2}(x - 2)^{\frac{1}{3}} - 1$

32) $f(x) = \dfrac{3}{16}(x + 2)^{\frac{4}{3}} - \dfrac{3}{2}(x + 2)^{\frac{1}{3}} + 2$

First Derivative Test

First Derivative Test: Suppose that c is a critical number of a continuous function f:

a) If f' changes from positive to negative at c, then f has a local maximum at c.

b) If f' changes from negative to positive at c, then f has a local minimum at c.

c) If f' does not change signs at c, then f has no local maximum or minimum at c.

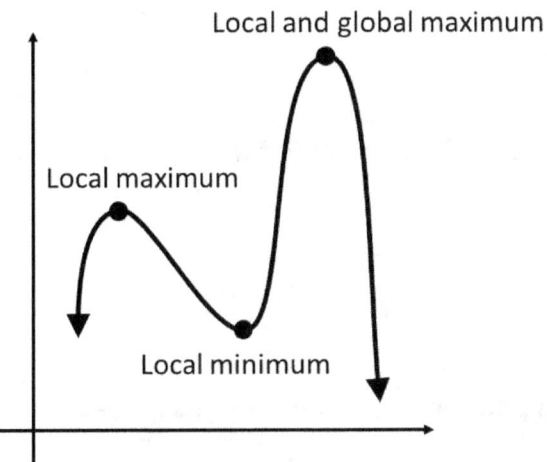

Concavity and Inflection Points

Concavity measures the rate of change of a function's derivative. Therefore, concavity is measured by the second derivative. A function is concave up when the derivative is increasing and is concave down when the derivative is decreasing. Concave up means it bends up, visually looking like a valley. Concave down means it bends down, looking like a hill. **Concave UP Like a CUP and Concave DOWN Like a FROWN**. An inflection point is defined at the point where the concavity changes sign. The second derivative at an inflection point will be zero and the second derivative will have opposite signs on each side of the inflection point.

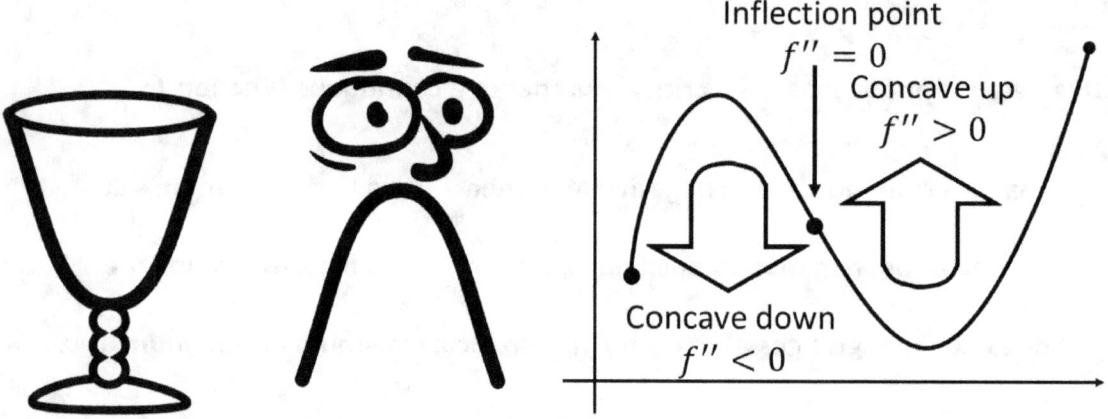

Second Derivative Test

Second Derivative Test: Suppose that $f'(c) = 0$ of a twice-differentiable function f:

a) If $f''(c) < 0$ then f has a local maximum at $x = c$.

b) If $f''(c) > 0$ then f has a local minimum at $x = c$.

c) If $f''(c) = 0$ then f may have an inflection point at $x = c$. **Check signs on each side.**

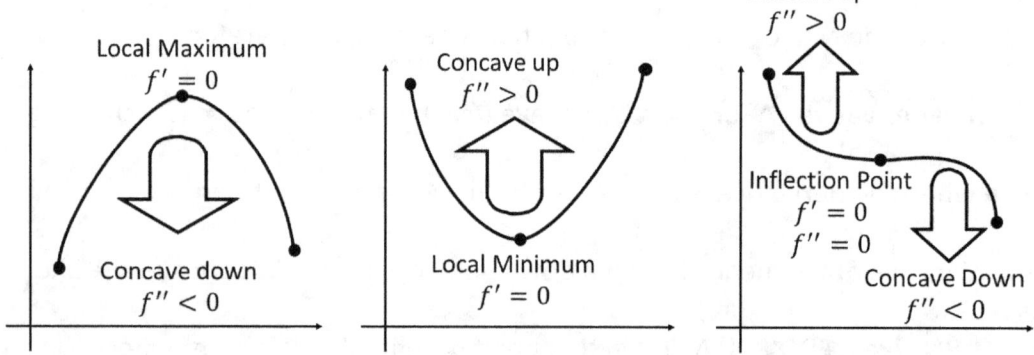

Curve Sketching

Using all the tools of the first derivative, second derivative, and algebra we can analyze and sketch many types of functions without much need of a calculator.

Ex. 1 Let $f(x) = \frac{x^3}{6} - \frac{x^2}{3} - \frac{2x}{3}$ Find the x and y intercepts, asymptotes, x-coordinates of the critical points, open intervals where the function is increasing and decreasing, x-coordinates of the inflection points, open intervals where the function is concave up and concave down, and relative minima and maxima. Using this information, sketch the graph of the function.

Solution: The easiest thing to find first is the y-intercept which is given by taking $f(0) = \frac{0^3}{6} - \frac{0^2}{3} - \frac{2(0)}{3} = 0$. Notice for polynomials the y-intercept will always be the constant. Here the y-intercept is the point (0, 0). Next, we will find the x-intercepts by setting the function to zero and solving: $\frac{x^3}{6} - \frac{x^2}{3} - \frac{2x}{3} = 0$ which can be factored as: $\frac{x}{6}(x^2 - 2x - 4) = 0$. This gives one solution of (0, 0). We cannot factor $x^2 - 2x - 4$, so we use the quadratic formula: $x = \frac{-(-2) \pm \sqrt{(-2)^2 - 4(1)(-4)}}{2(1)} = 1 \pm \sqrt{5} \approx 3.2361 \text{ or } -1.2361$. So the three x-intercepts are $(1 - \sqrt{5}, 0), (0,0), (1 + \sqrt{5}, 0)$. There are no asymptotes for a polynomial function. Next, we find the critical points by taking the derivative: $f'(x) = \frac{x^2}{2} - \frac{2x}{3} - \frac{2}{3}$. Set to zero and solve:

$\frac{x^2}{2} - \frac{2x}{3} - \frac{2}{3} = 0$ Factoring gives: $\frac{1}{6}(x - 2)(3x + 2) = 0$. This gives critical values of $x = 2, -\frac{2}{3}$.

Test the intervals for increasing and decreasing. $f'(-10) = \frac{(-10)^2}{2} - \frac{2(-10)}{3} - \frac{2}{3} = 56$ is positive.

$f'(0) = \frac{(0)^2}{2} - \frac{2(0)}{3} - \frac{2}{3} = -\frac{2}{3}$ is negative. $f'(10) = \frac{(10)^2}{2} - \frac{2(10)}{3} - \frac{2}{3} = \frac{128}{3}$ is positive.

Therefore, the intervals of increasing are $(-\infty, -\frac{2}{3}) \cup (2, \infty)$. And the intervals of decreasing are $(-\frac{2}{3}, 2)$. Next, we find any possible inflection points and check for concavity by taking the second derivative: $f''(x) = x - \frac{2}{3}$. When set to zero this gives $x = \frac{2}{3}$. Check concavity by evaluating the second derivative on each side: $f''(0) = 0 - \frac{2}{3} = -\frac{2}{3}$. Negative second derivative indicates concave down. Check the other side: $f''(10) = 10 - \frac{2}{3} = \frac{28}{3}$. Positive second derivative indicates concave up. Concave up: $(\frac{2}{3}, \infty)$ and concave down: $(-\infty, \frac{2}{3})$. Since it switches concavity there is an inflection point at $x = \frac{2}{3}$. Evaluate $f\left(\frac{2}{3}\right) = \frac{(2/3)^3}{6} - \frac{\left(\frac{2}{3}\right)^2}{3} - \frac{2\left(\frac{2}{3}\right)}{3} = -\frac{44}{81}$. The inflection point is $(\frac{2}{3}, -\frac{44}{81})$. The critical point at $x = -\frac{2}{3}$ is a relative maximum by either the first derivative test or the second derivative test. $f\left(-\frac{2}{3}\right) = \frac{(-2/3)^3}{6} - \frac{\left(-\frac{2}{3}\right)^2}{3} - \frac{2\left(-\frac{2}{3}\right)}{3} = \frac{20}{81}$ So, we have a relative maximum at $\left(-\frac{2}{3}, \frac{20}{81}\right)$. The critical point at $x = 2$ is a relative minimum by either the first derivative test or the second derivative test. $f(2) = \frac{(2)^3}{6} - \frac{(2)^2}{3} - \frac{2(2)}{3} = -\frac{4}{3}$ This gives the relative minimum at $\left(2, -\frac{4}{3}\right)$. We use this all this information in sketching the graph. When sketching the curve, it is not necessary to match up the exact shape of the curve, just the salient features that we identified: intercepts, extrema, increasing/decreasing behavior, inflection, and concavity.

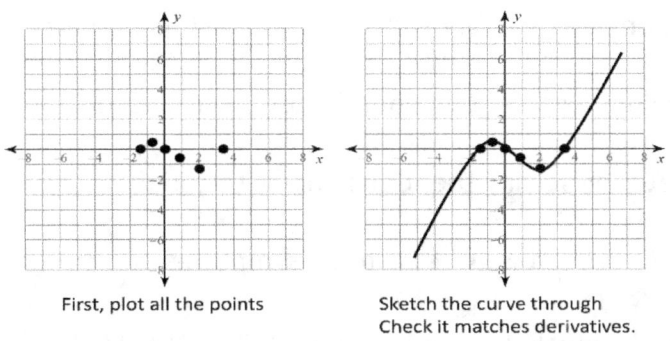

First, plot all the points

Sketch the curve through
Check it matches derivatives.

Here's the full solution:

$y = \frac{x^3}{6} - \frac{x^2}{3} - \frac{2x}{3}$

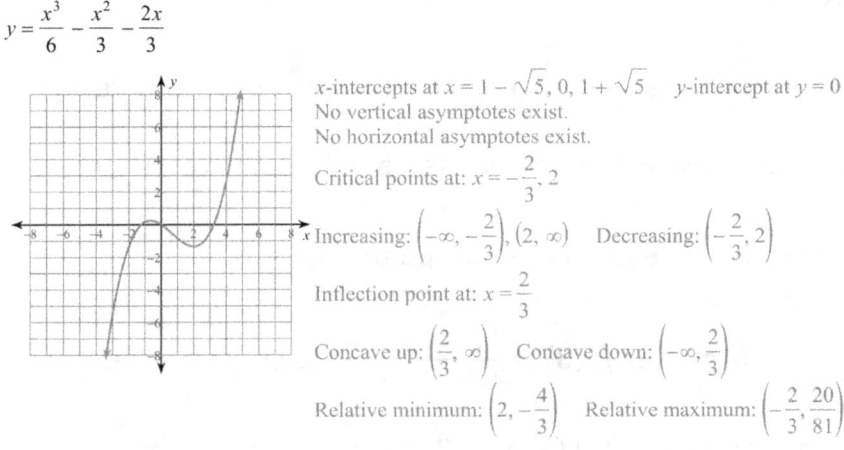

x-intercepts at $x = 1 - \sqrt{5}, 0, 1 + \sqrt{5}$ y-intercept at $y = 0$
No vertical asymptotes exist.
No horizontal asymptotes exist.
Critical points at: $x = -\frac{2}{3}, 2$
Increasing: $\left(-\infty, -\frac{2}{3}\right), (2, \infty)$ Decreasing: $\left(-\frac{2}{3}, 2\right)$
Inflection point at: $x = \frac{2}{3}$
Concave up: $\left(\frac{2}{3}, \infty\right)$ Concave down: $\left(-\infty, \frac{2}{3}\right)$
Relative minimum: $\left(2, -\frac{4}{3}\right)$ Relative maximum: $\left(-\frac{2}{3}, \frac{20}{81}\right)$

Ex. 2 Let $f(x) = -\frac{x}{x^2-4}$ Find the x and y intercepts, asymptotes, x-coordinates of the critical points, open intervals where the function is increasing and decreasing, x-coordinates of the inflection points, open intervals where the function is concave up and concave down, and relative minima and maxima. Using this information, sketch the graph of the function.

<u>Solution:</u> The easiest thing to find first is the y-intercept which is given by taking $f(0) = -\frac{0}{0^2-4} = 0$. The y-intercept is the point (0, 0). Next, we will find the x-intercepts by setting the function to zero and solving: $\frac{-x}{x^2-4} = 0$ A rational function is zero when the numerator is zero,

which in this case: $x = 0$. This gives one solution of $(0, 0)$. Since this is a rational function we need to check for asymptotes. If we factor the numerator and denominator and cancel out common factors, we can find vertical asymptotes by setting the remaining factors in the denominator to zero. $x^2 - 4 = (x+2)(x-2) = 0$. This gives vertical asymptotes at $x = -2$ and $x = 2$. Horizontal asymptotes can be found by taking the limit as x goes to positive or negative infinity. On both sides this gives a horizontal asymptote of $y = 0$. Next, we find the critical points by taking the derivative of: $f(x) = -x(x^2 - 4)^{-1}$. $f'(x) = \frac{x^2 + 4}{(x^2 - 4)^2}$. Set the numerator to zero and solve: $x^2 + 4 = 0$. This does not have any real solutions, so there are no critical points. Even though we don't have any critical points we still need to check for intervals of increasing and decreasing around each of the asymptotes. $f'(-3) = \frac{(-3)^2 + 4}{((-3)^2 - 4)^2} = \frac{13}{25}$ is positive. $f'(0) = \frac{(0)^2 + 4}{((0)^2 - 4)^2} = \frac{1}{4}$ is negative. $f'(3) = \frac{(3)^2 + 4}{((3)^2 - 4)^2} = \frac{13}{25}$ is positive. Therefore, the intervals of increasing are $(-\infty, -2) \cup (-2, 2) \cup (2, \infty)$. There are no intervals of decreasing. Next, we find any possible inflection points and check for concavity by taking the second derivative: $f''(x) = -\frac{2x(x^2 + 12)}{(x^2 - 4)^3}$. When the numerator is set to zero, we only get one real solution at $x = 0$. Check concavity by evaluating the second derivative on each side: $f''(-1) = -\frac{2(-1)((-1)^2 + 12)}{((-1)^2 - 4)^3} = -\frac{26}{27}$. Negative second derivative indicates concave down. Check the other side: $f''(1) = -\frac{2(1)(1^2 + 12)}{((1)^2 - 4)^3} = \frac{26}{27}$. Positive second derivative indicates concave up. Concave up: $(0, \infty)$ and concave down: $(-\infty, 0)$. Since it switches concavity there is an inflection point at $x = 0$. We already know what the y-value is at 0. So the inflection point is

(0,0). Since there are no critical points, we know there are no relative extrema for this graph.

We then use this all this information in sketching the graph.

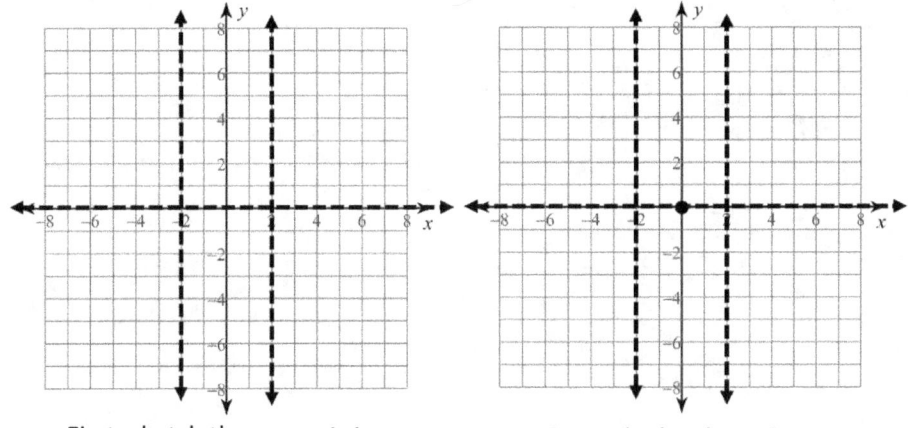

First, sketch the asymptotes

Second, plot the points

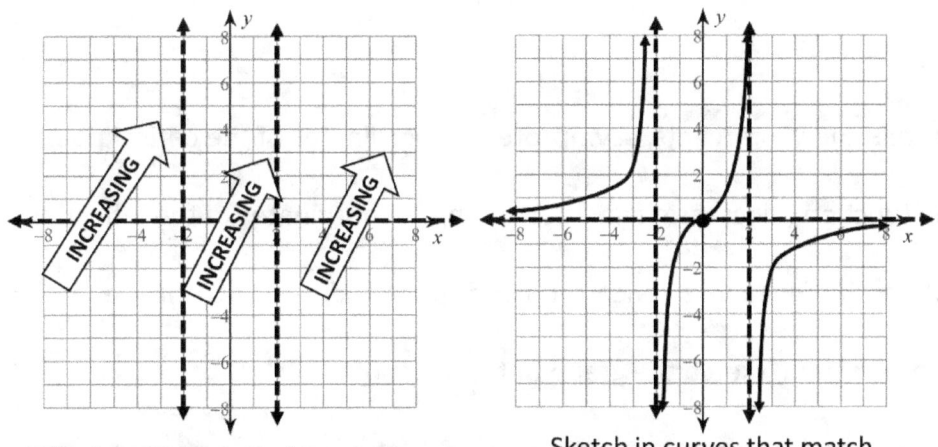

Third, use increasing/decreasing

Sketch in curves that match
Notice it only crosses x-axis at 0
Curves will bend with the asymptotes
Also check concavity

Here's the full solution:

$$y = -\frac{x}{x^2 - 4}$$

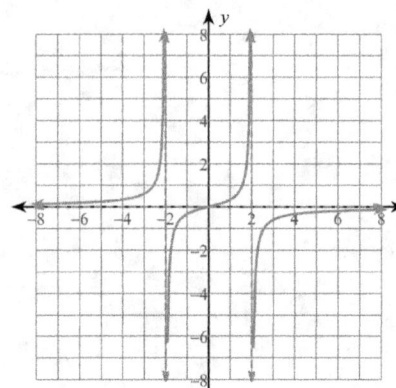

x-intercept at $x = 0$ y-intercept at $y = 0$
Vertical asymptotes at: $x = -2, 2$
Horizontal asymptote at: $y = 0$
No critical points exist.
Increasing: $(-\infty, -2), (-2, 2), (2, \infty)$ Decreasing: No intervals exist.
Inflection point at: $x = 0$
Concave up: $(-\infty, -2), (0, 2)$ Concave down: $(-2, 0), (2, \infty)$
No relative minima.   No relative maxima.

Practice 30

For each problem, find the: x and y intercepts, asymptotes, x-coordinates of the critical points, open intervals where the function is increasing and decreasing, x-coordinates of the inflection points, open intervals where the function is concave up and concave down, and relative minima and maxima. Using this information, sketch the graph of the function.

1) $y = \dfrac{x^2}{2} - 2x$

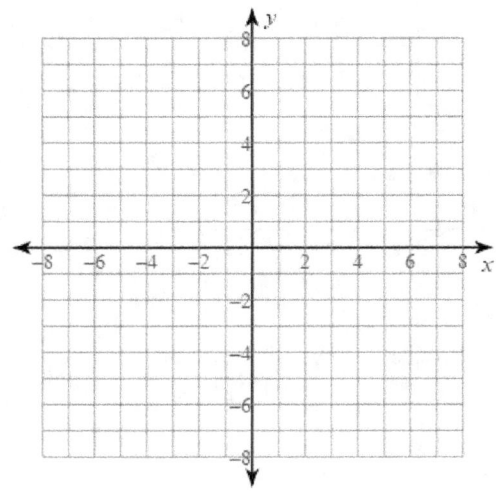

2) $y = x^2 - 4x - 1$

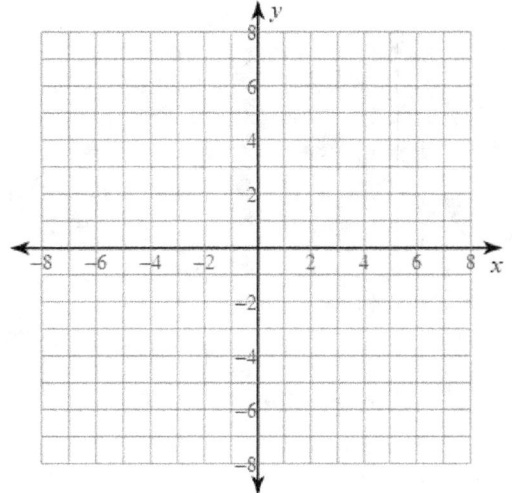

3) $y = -x^2 - 4x - 2$

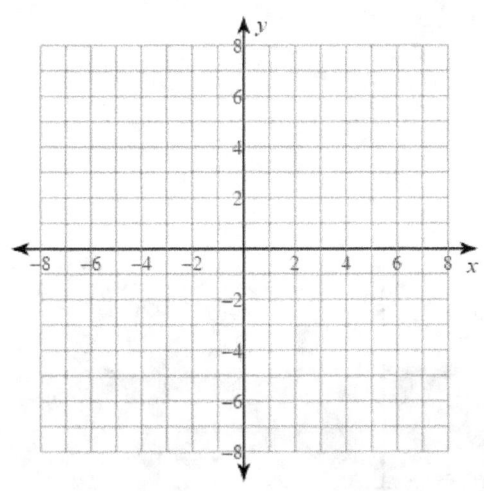

4) $y = \dfrac{x^2}{2} - 4x + 2$

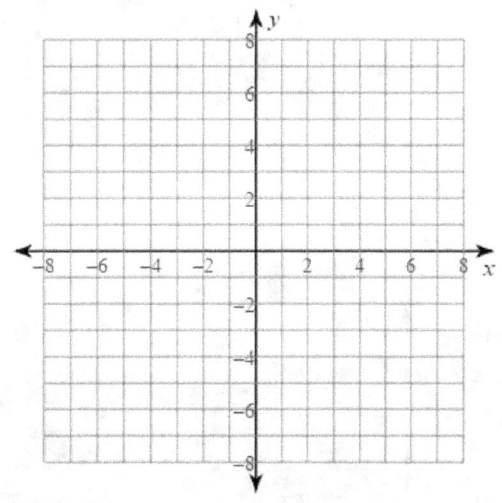

5) $y = 2x^3 - 4x^2$

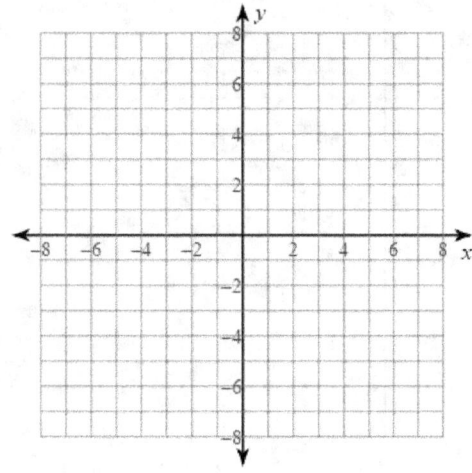

6) $y = \dfrac{x^3}{6} + \dfrac{x^2}{3} - \dfrac{2x}{3}$

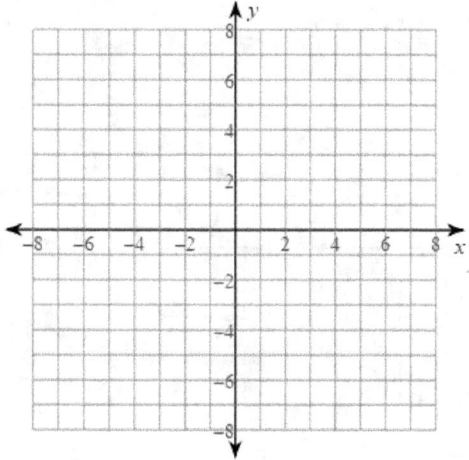

7) $y = -\dfrac{x^4}{8} + \dfrac{x^2}{4} - \dfrac{1}{8}$

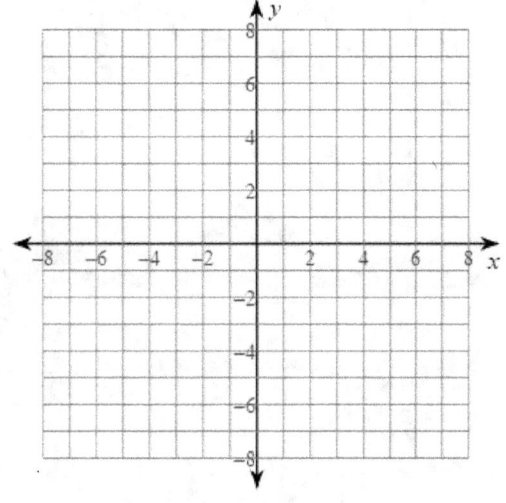

8) $y = x^4 - 3x^2$

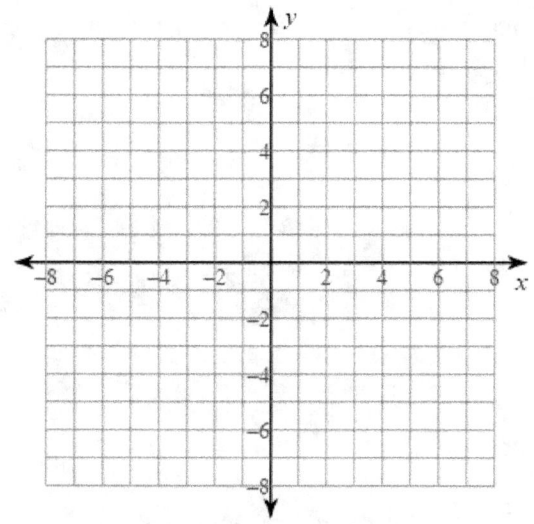

9) $y = -\dfrac{x^2}{3x-3}$

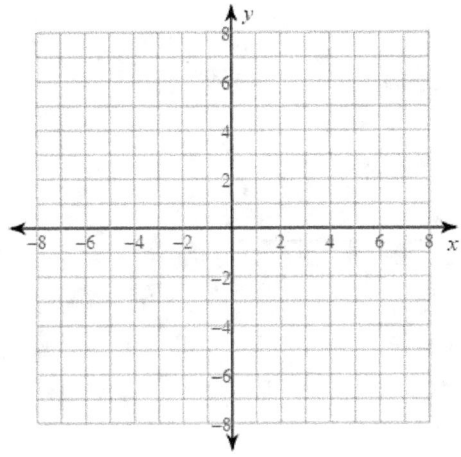

10) $y = \dfrac{3x}{x-3}$

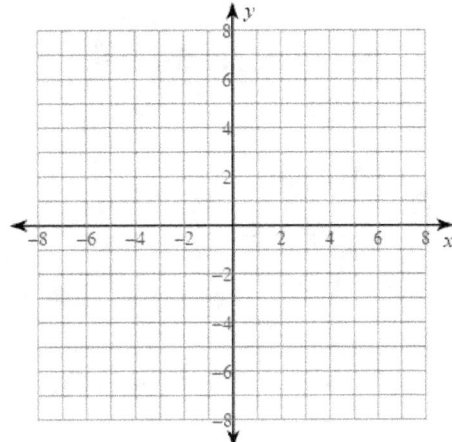

11) $y = -\dfrac{1}{x+3}$

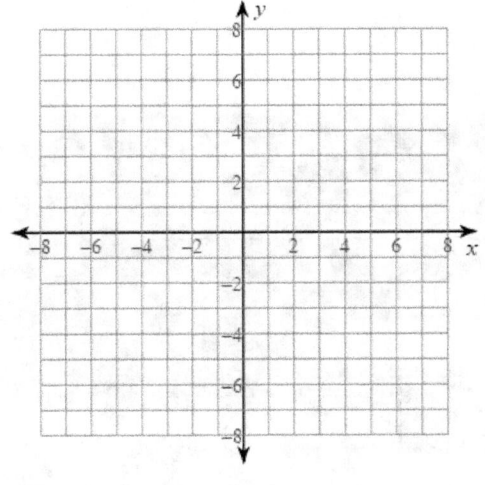

12) $y = \dfrac{x^2}{2x+2}$

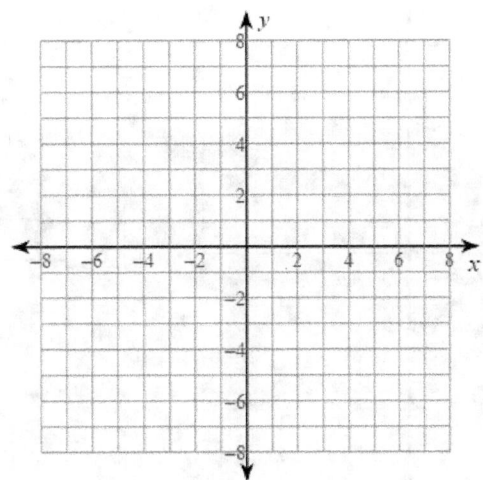

13) $y = \dfrac{x^2 - 1}{x^3}$

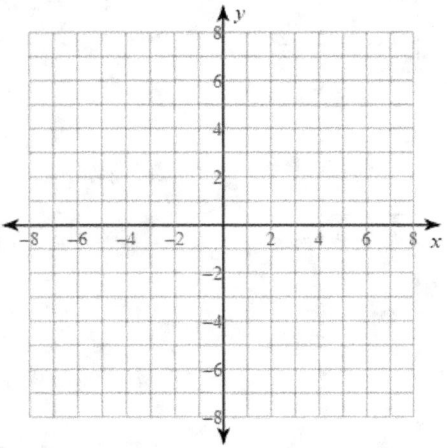

14) $y = \dfrac{1}{x^2 - 16}$

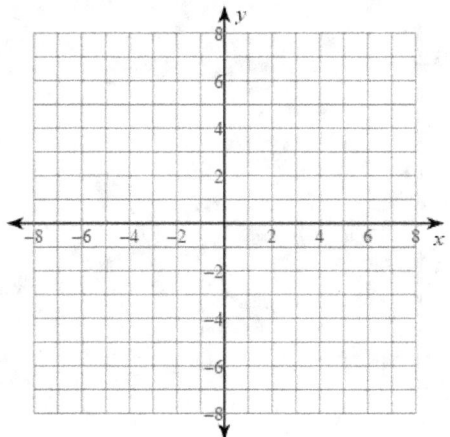

15) $y = \dfrac{3}{x^2 + 3}$

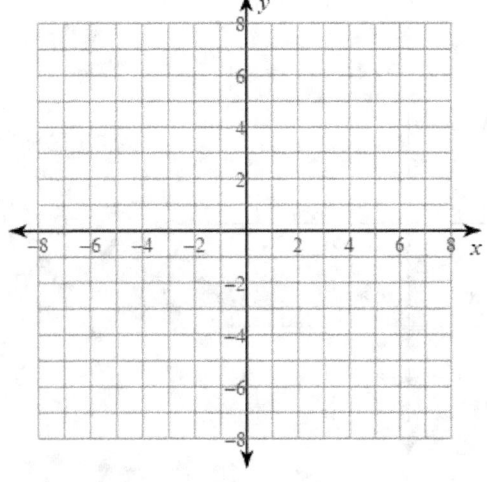

16) $y = \dfrac{x^3}{x^2 - 4}$

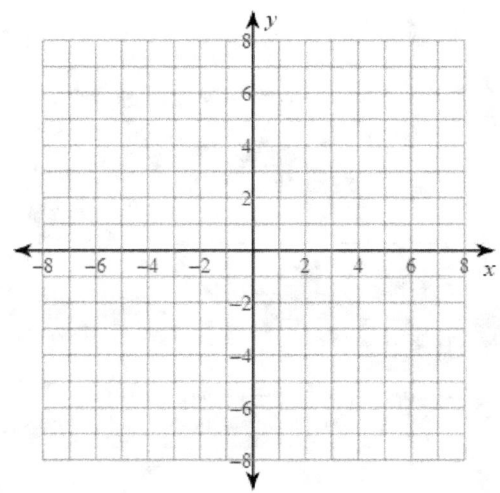

17) $y = (3x-18)^{\frac{2}{3}}$

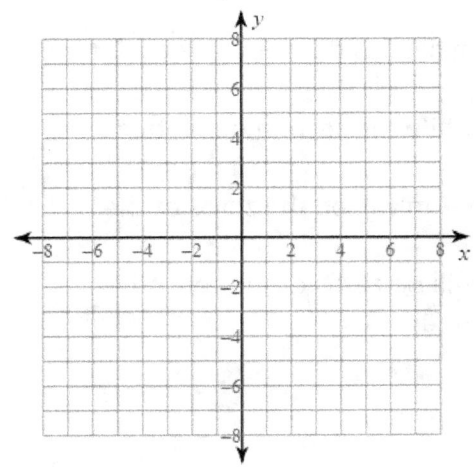

18) $y = -(-3x+3)^{\frac{1}{2}}$

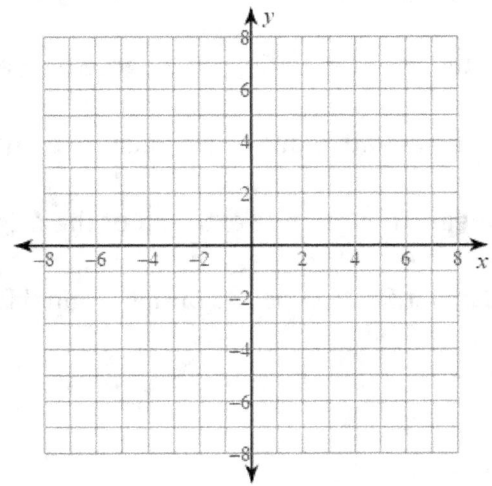

19) $y = -\frac{3}{16}(x+2)^{\frac{4}{3}} + \frac{3}{2}(x+2)^{\frac{1}{3}}$

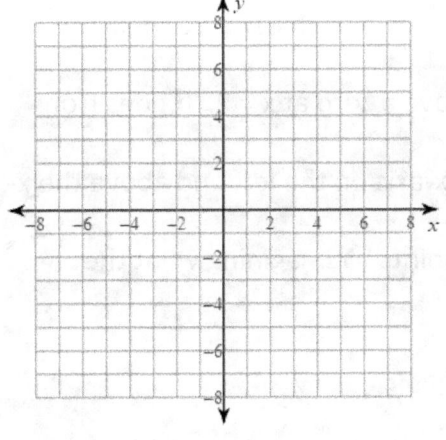

20) $y = -\frac{1}{5}(x-4)^{\frac{5}{3}} - 2(x-4)^{\frac{2}{3}}$

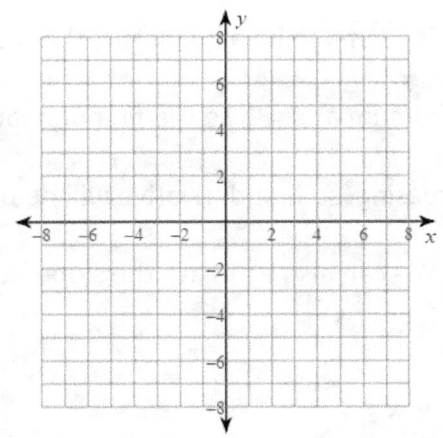

Relationship among $f(x)$, $f'(x)$, and $f''(x)$

With polynomials, the relationship between $f(x)$, $f'(x)$, and $f''(x)$ is easier to see because as we take the derivative the degree of the polynomial goes down by one each time. For example, the derivative of a cubic must be a quadratic, while the second derivative of a cubic will be linear. The relative extrema of the original function will be the zeros of the derivative and any inflection points of the original function will be zeros of the second derivative function.

Ex. 2

Sketch $f'(x)$ and $f''(x)$

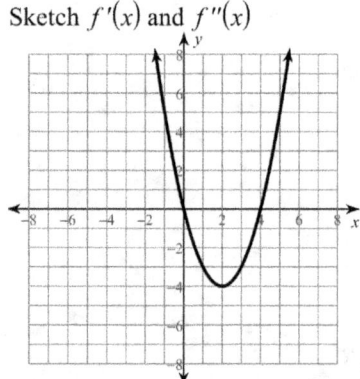

Solution: Since this seems to be a quadratic, the first derivative will be linear. It appears to have a relative minimum at x = 2, so the first derivative will have a zero at x = 2. It goes from decreasing to increasing, so first derivative will be below the x-axis on the left and above the x-axis on the right. We can also estimate the slope at various points by imagining what the tangent line is.

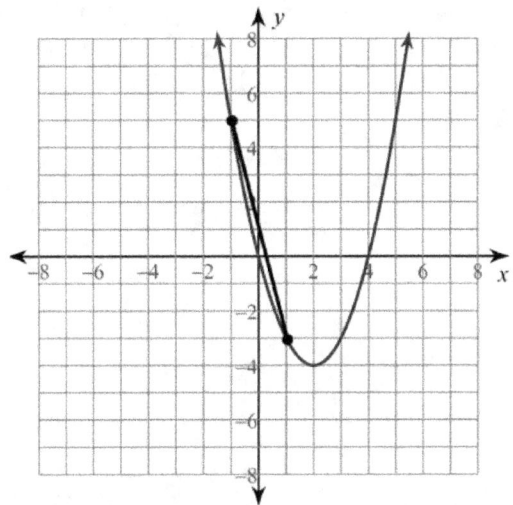

We can use the points $(-1, 5)$ and $(1, -3)$ to estimate that the slope is $\frac{-3-5}{1--1} = -\frac{8}{2} = -4$

For example, at the point (0, 0), the slope appears to be close to – 4. This tells us that (0, – 4) will be a y-intercept on the first derivative function.

Here's the graph of $f'(x)$:

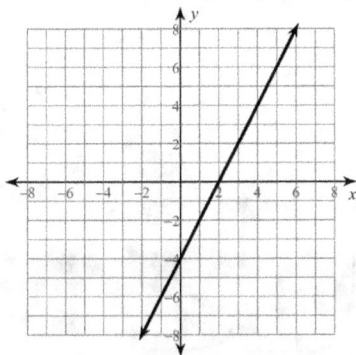

From this we can deduce what the second derivative looks like. Since this graph is linear, the second derivative will be constant. The slope of this line will give us the value of the second derivative $f''(x) = 2$:

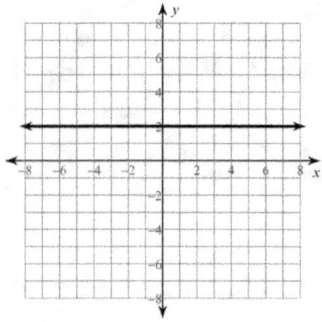

Here's another example with a cubic:

$f(x)$ 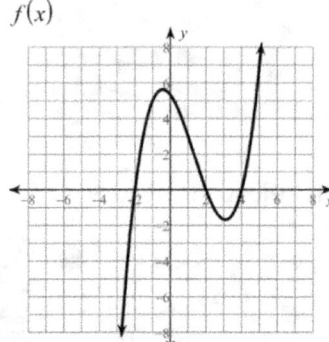 $f'(x)$ 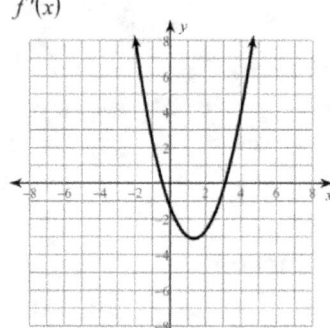 $f''(x)$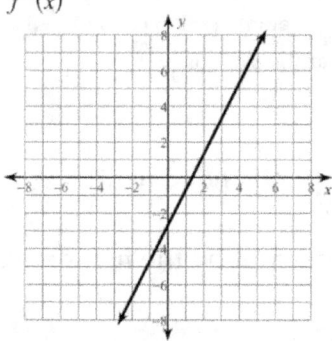

With rational functions it also helps to make note of intervals of increasing/decreasing and concavity:

$f(x)$ 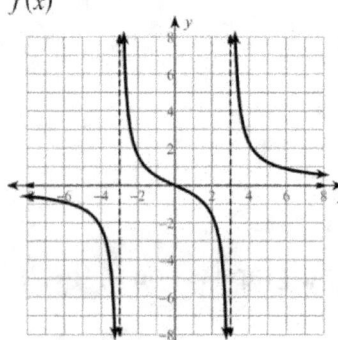 $f'(x)$ 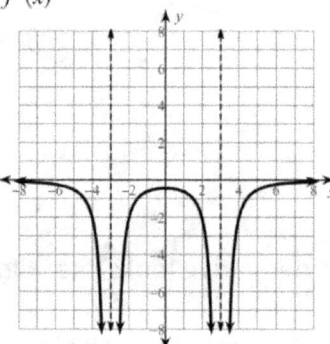 $f''(x)$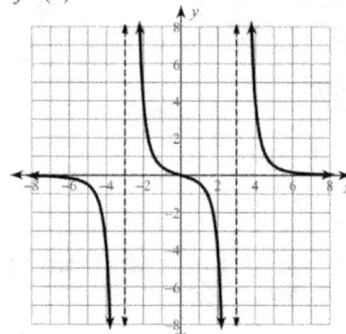

Practice 31

Given the graph of $f(x)$, sketch an approximate graph of $f'(x)$.

1)

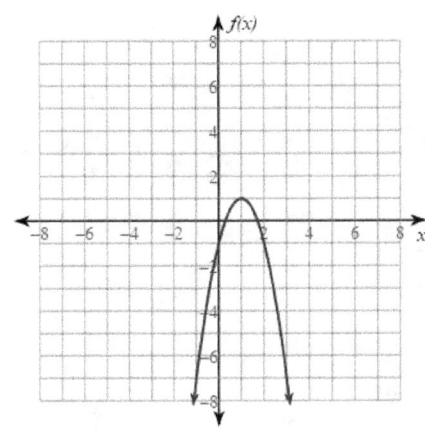 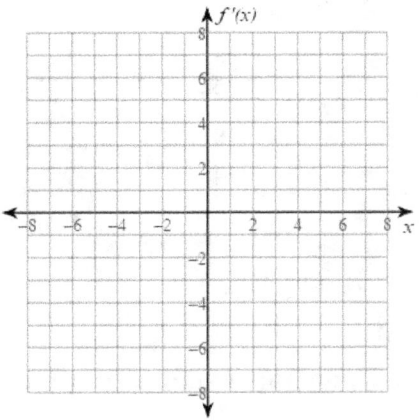

2)

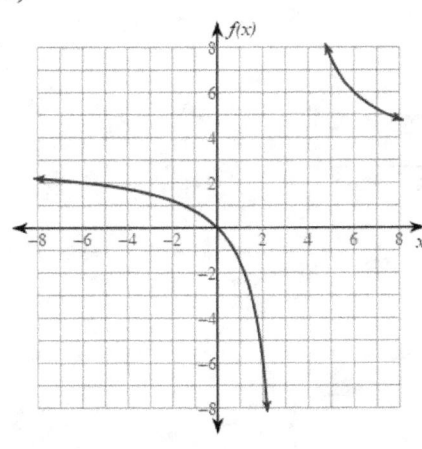 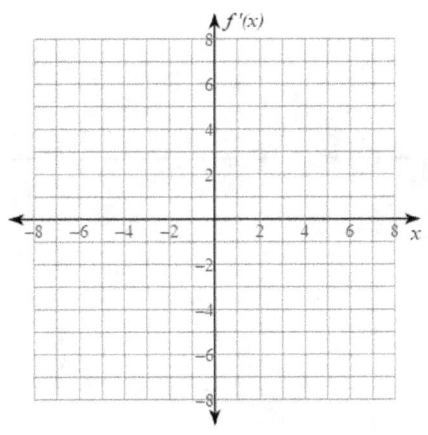

3)

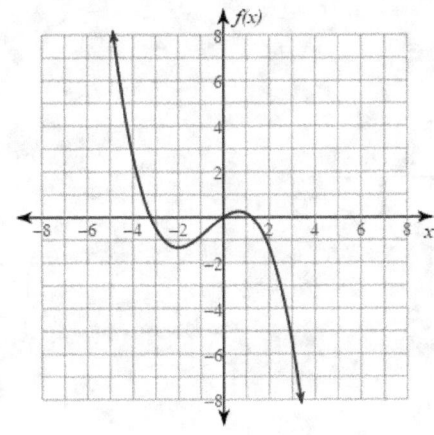 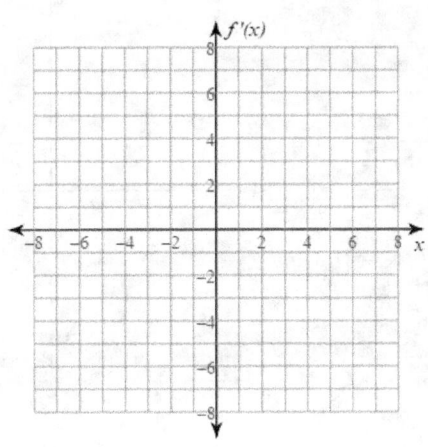

4)

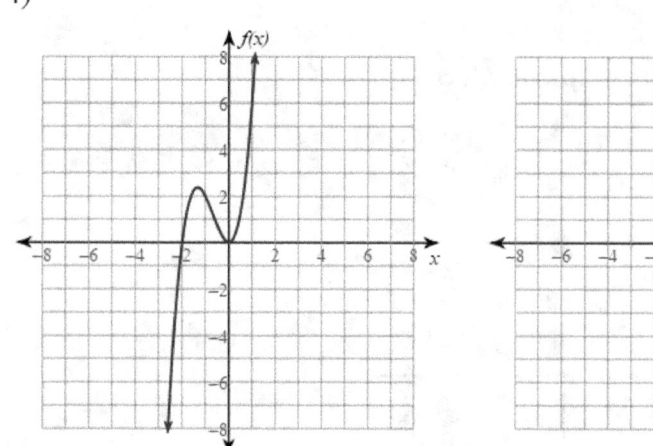

Given the graph of $f(x)$, sketch an approximate graph of $f''(x)$.

5)

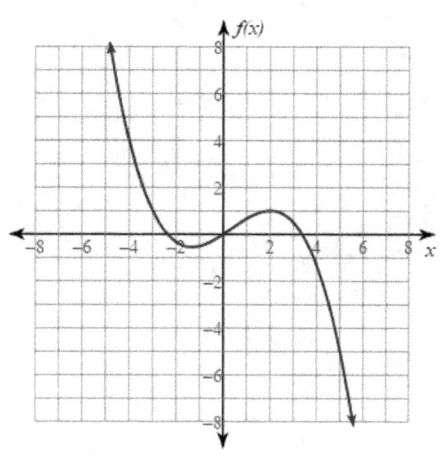

6)

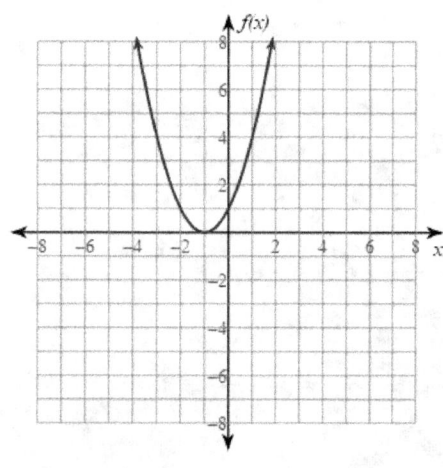
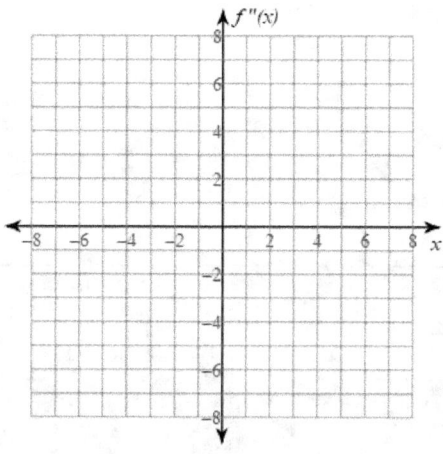

7)

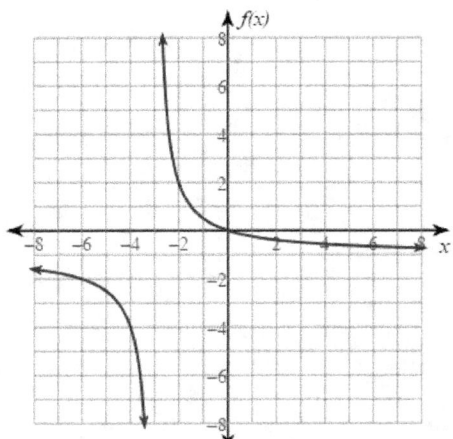

8)

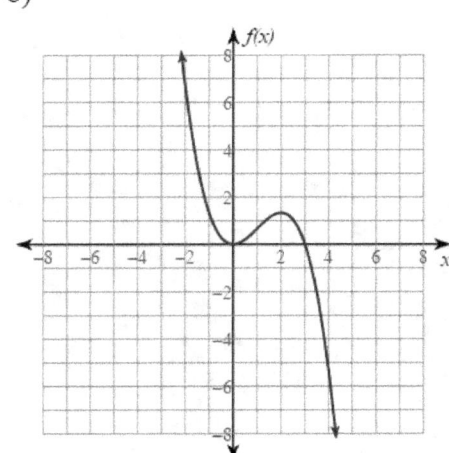
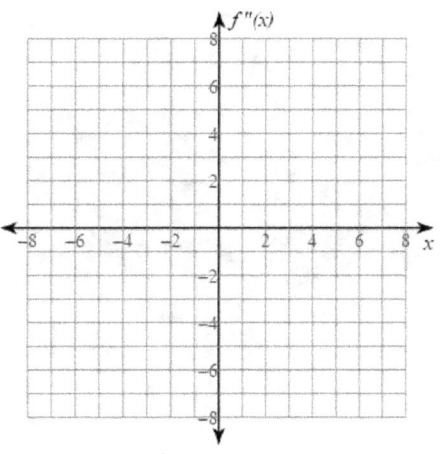

Given the graph of $f(x)$, sketch an approximate graph of $f'(x)$.

9)

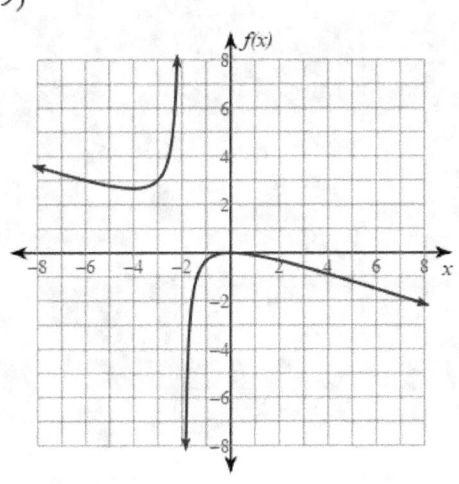
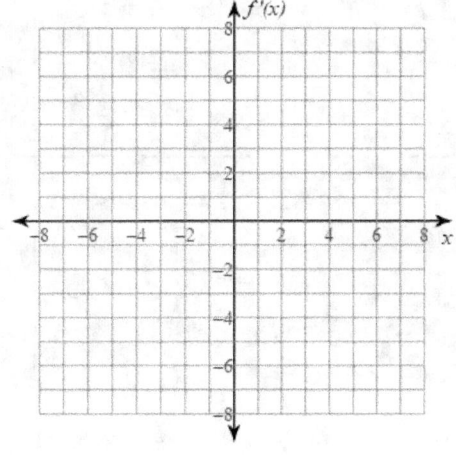

10)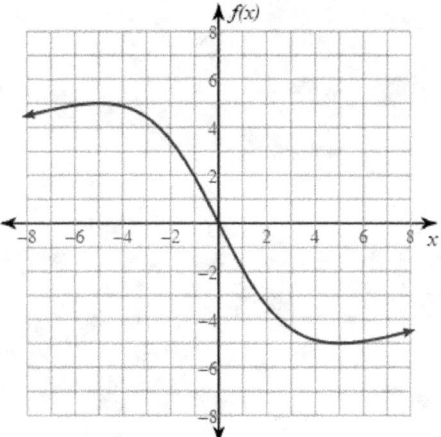

11)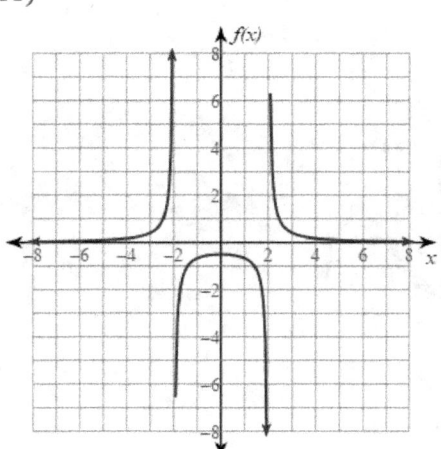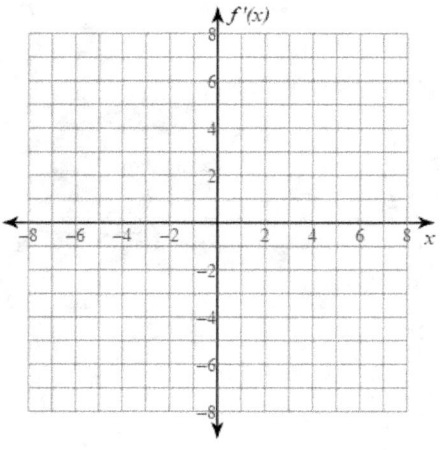

12)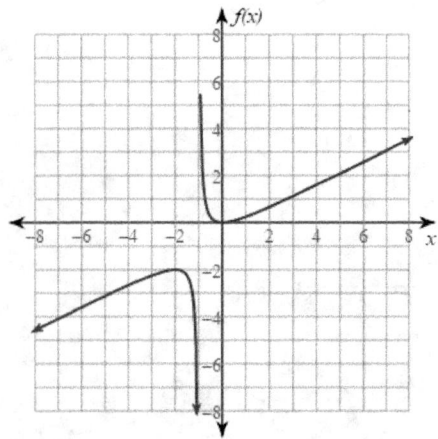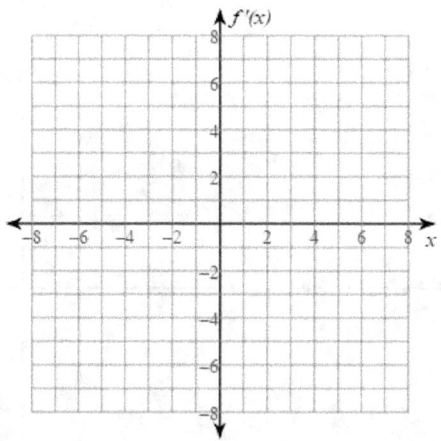

Given the graph of $f(x)$, sketch an approximate graph of $f''(x)$.

13)

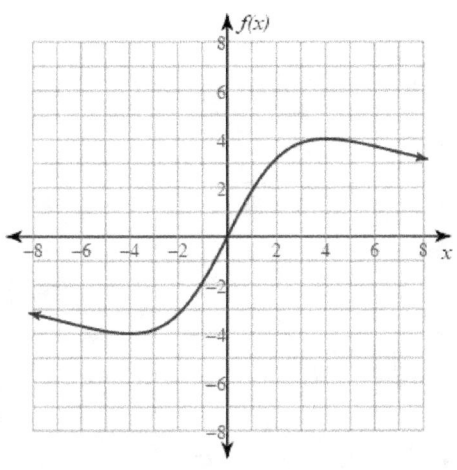 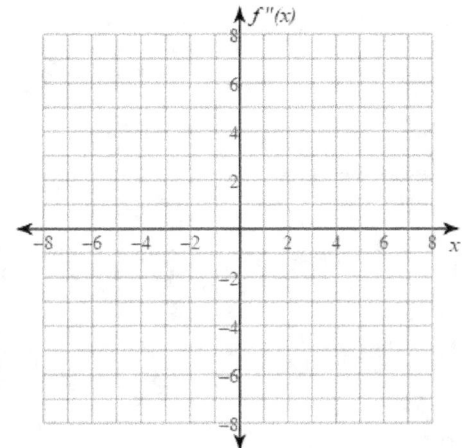

14)

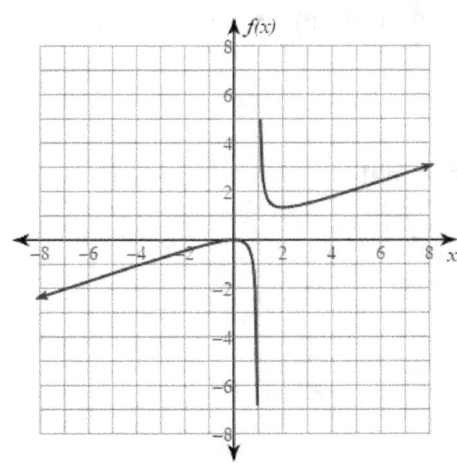 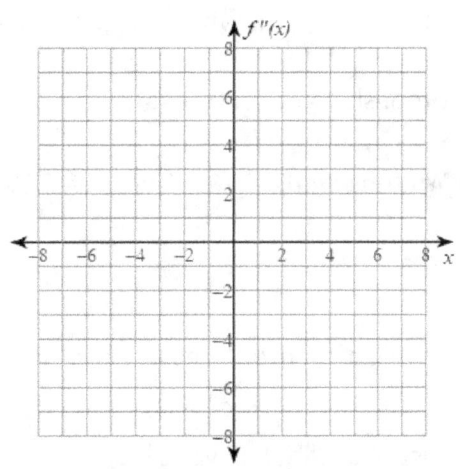

15)

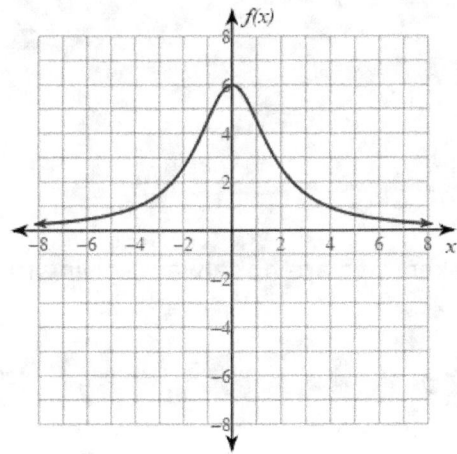 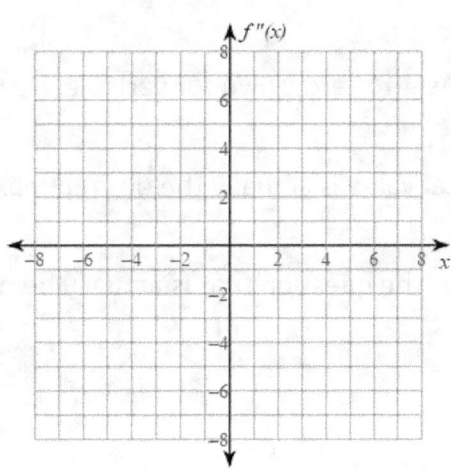

16)

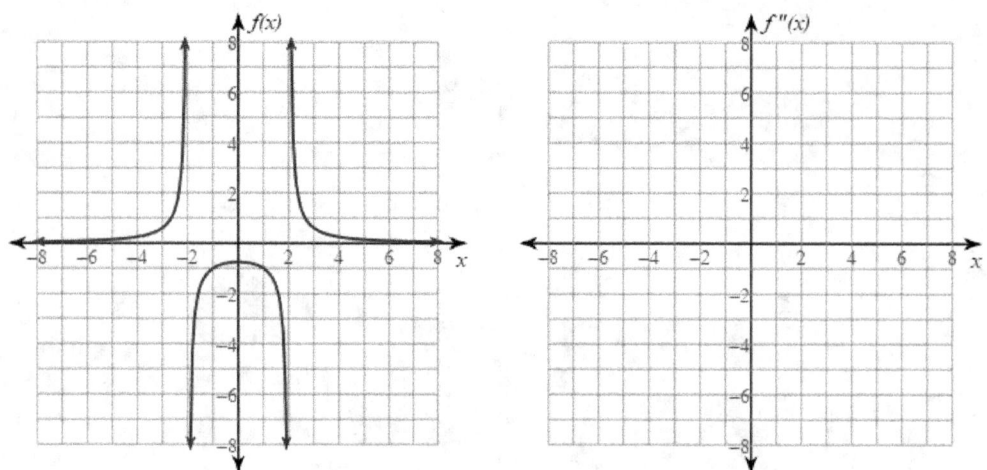

Optimization Problems

Optimization problems happen when we try to maximize or minimize a particular variable in order to accomplish a goal. For example, we may want to maximize profit when selling merchandise or we may want to minimize the amount of material we need when building something.

Here are the steps to optimization problems in calculus:

1. Draw a picture if applicable. Label all variables.

2. Write an equation to maximize/minimize.

3. Take the derivative, set to zero, solve for variable.

4. Make sure the critical value is actually the desired max/min.

5. Make sure to answer the question that is actually being asked...we are not always solving for the variable

Ex. 1 A rancher wants to construct two identical rectangular corrals using 500 ft of fencing. The rancher decides to build them adjacent to each other, so they share fencing on one side. What dimensions should the rancher use to construct each corral so that together, they will enclose the largest possible area?

Solution: First step, draw a picture and label it:

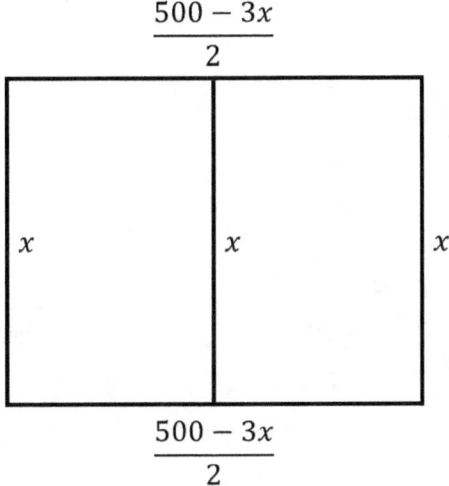

We label one side x, and then use that to solve for this missing dimension. Since the perimeter is 500, the other side is 500 minus 3x from the 3 sides, divided by 2. We want to minimize the area which we can express: $A = lw = x\left(\frac{500-3x}{2}\right) = 250x - \frac{3}{2}x^2$. Take the derivative: $A'(x) = 250 - 3x$. Set to zero, $250 - 3x = 0$. $x = \frac{250}{3}$.

Since the second derivative is $A''(x) = -3$, so the function is always concave down, therefore it will have a max at the critical value. Graphing the function will also confirm this. To find the other side plug the value into $\frac{500-3(250/3)}{2} = 125$. *This means that the sides of each corral are* $\frac{250}{3}$ *by* $\frac{125}{2}$.

Ex. 2 A graphic designer is asked to create a movie poster with a 72 in² photo surrounded by a 2 inch border at the top and bottom and a 1 inch border on each side. What overall dimensions for the poster should the designer choose to use the least amount of paper?

Solution: First step, draw a picture and label it:

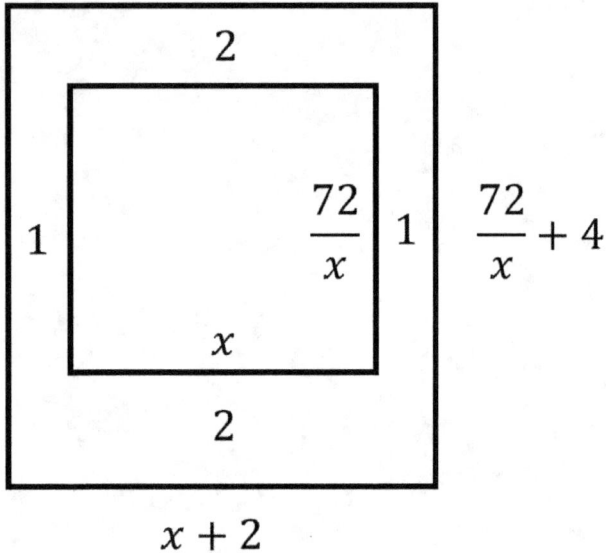

We label one side of the photo x, and since the area is 72, the other side is $\frac{72}{x}$. The sides of the poster will then be $x + 2$ and $\frac{72}{x} + 4$. We want to minimize the area which we can express:

$A = lw = (x + 2)\left(\frac{72}{x} + 4\right) = 72 + 4x + \frac{144}{x} + 8 = 4x + \frac{144}{x} + 80$. Take the derivative:

$A'(x) = 4 - \frac{144}{x^2}$. Set to zero: $4 - \frac{144}{x^2} = 0 \Rightarrow 4 = \frac{144}{x^2} \Rightarrow 4x^2 = 144 \Rightarrow x^2 = 36 \Rightarrow x = 6$. We take only the positive solution since we are solving for a distance. We need to solve for the dimensions of the outside poster, so we substitute 6 into $x + 2$ and $\frac{72}{x} + 4$ to give the dimensions of *8 inches by 16 inches*.

Practice 32

Solve each optimization problem

1) A rancher wants to construct two identical rectangular corrals using 200 ft of fencing. The rancher decides to build them adjacent to each other, so they share fencing on one side. What dimensions should the rancher use to construct each corral so that together, they will enclose the largest possible area?

2) A rancher wants to construct two identical rectangular corrals using 400 ft of fencing. The rancher decides to build them adjacent to each other, so they share fencing on one side. What dimensions should the rancher use to construct each corral so that together, they will enclose the largest possible area?

3) A farmer wants to construct a rectangular pigpen using 400 ft of fencing. The pen will be built next to an existing stone wall, so only three sides of fencing need to be constructed to enclose the pen. What dimensions should the farmer use to construct the pen with the largest possible area?

4) A farmer wants to construct a rectangular pigpen using 100 ft of fencing. The pen will be built next to an existing stone wall, so only three sides of fencing need to be constructed to enclose the pen. What dimensions should the farmer use to construct the pen with the largest possible area?

5) A cryptography expert is deciphering a computer code. To do this, the expert needs to minimize the product of a positive rational number and a negative rational number, given that the positive number is exactly 5 greater than the negative number. What final product is the expert looking for?

6) A cryptography expert is deciphering a computer code. To do this, the expert needs to minimize the product of a positive rational number and a negative rational number, given that the positive number is exactly 8 greater than the negative number. What final product is the expert looking for?

7) A company has started selling a new type of smartphone at the price of $130 - 0.1x$ where x is the number of smartphones manufactured per day. The parts for each smartphone cost $60 and the labor and overhead for running the plant cost $4000 per day. How many smartphones should the company manufacture and sell per day to maximize profit?

8) A company has started selling a new type of smartphone at the price of $140 - 0.05x$ where x is the number of smartphones manufactured per day. The parts for each smartphone cost $70 and the labor and overhead for running the plant cost $7000 per day. How many smartphones should the company manufacture and sell per day to maximize profit?

9) A supermarket employee wants to construct an open-top box from a 10 by 16 in piece of cardboard. To do this, the employee plans to cut out squares of equal size from the four corners so the four sides can be bent upwards. What size should the squares be in order to create a box with the largest possible volume?

10) A supermarket employee wants to construct an open-top box from a 14 by 30 in piece of cardboard. To do this, the employee plans to cut out squares of equal size from the four corners so the four sides can be bent upwards. What size should the squares be in order to create a box with the largest possible volume?

11) A graphic designer is asked to create a movie poster with a 50 in^2 photo surrounded by a 4 in border at the top and bottom and a 2 in border on each side. What overall dimensions for the poster should the designer choose to use the least amount of paper?

12) A graphic designer is asked to create a movie poster with a 98 in^2 photo surrounded by a 2 in border at the top and bottom and a 1 in border on each side. What overall dimensions for the poster should the designer choose to use the least amount of paper?

13) Engineers are designing a box-shaped aquarium with a square bottom and an open top. The aquarium must hold 1372 ft³ of water. What dimensions should they use to create an acceptable aquarium with the least amount of glass?

14) Engineers are designing a box-shaped aquarium with a square bottom and an open top. The aquarium must hold 2048 ft³ of water. What dimensions should they use to create an acceptable aquarium with the least amount of glass?

15) A geometry student wants to draw a rectangle inscribed in the ellipse $x^2 + 4y^2 = 49$. What is the area of the largest rectangle that the student can draw?

16) A geometry student wants to draw a rectangle inscribed in the ellipse $x^2 + 4y^2 = 36$. What is the area of the largest rectangle that the student can draw?

17) A geometry student wants to draw a rectangle inscribed in a semicircle of radius 3. If one side must be on the semicircle's diameter, what is the area of the largest rectangle that the student can draw?

18) A geometry student wants to draw a rectangle inscribed in a semicircle of radius 8. If one side must be on the semicircle's diameter, what is the area of the largest rectangle that the student can draw?

19) Which points on the graph of $y = 5 - x^2$ are closest to the point $(0, 1)$?

20) Which points on the graph of $y = 4 - x^2$ are closest to the point $(0, 3)$?

21) Which point on the graph of $y = \sqrt{x}$ is closest to the point $(1, 0)$?

22) Which point on the graph of $y = \sqrt{x}$ is closest to the point $(5, 0)$?

23) Two vertical poles, one 5 ft high and the other 10 ft high, stand 36 feet apart on a flat field. A worker wants to support both poles by running rope from the ground to the top of each post. If the worker wants to stake both ropes in the ground at the same point, where should the stake be placed to use the least amount of rope?

24) Two vertical poles, one 12 ft high and the other 36 ft high, stand 20 feet apart on a flat field. A worker wants to support both poles by running rope from the ground to the top of each post. If the worker wants to stake both ropes in the ground at the same point, where should the stake be placed to use the least amount of rope?

25) An architect is designing a composite window by attaching a semicircular window on top of a rectangular window, so the diameter of the top window is equal to and aligned with the width of the bottom window. If the architect wants the perimeter of the composite window to be 12 ft, what dimensions should the bottom window be in order to create the composite window with the largest area?

26) An architect is designing a composite window by attaching a semicircular window on top of a rectangular window, so the diameter of the top window is equal to and aligned with the width of the bottom window. If the architect wants the perimeter of the composite window to be 10 ft, what dimensions should the bottom window be in order to create the composite window with the largest area?

Integration and Accumulation of Change

So far, we have talked about limits and derivatives. The idea of a limit allows us to make calculations for change as something becomes infinitesimally small or infinitely large. The concept of a derivative allows us to find instantaneous rate of change or the slope of a tangent at any particular point along a curve. Now we move to the next big concept of calculus: the integral. Integration or taking an integral is actual a sum of infinitesimal slices. Integration can be used to find the area under a curve. Differentiation breaks up a function into tiny bits, integration put these little bits into a complete whole. Integration is the inverse of differentiation. Sometimes taking an integral is also called finding the antiderivative.

Accumulation of Change

The integration gives us the area under a curve. The integral sign is $\int$ which is an elongated S representing a sum.

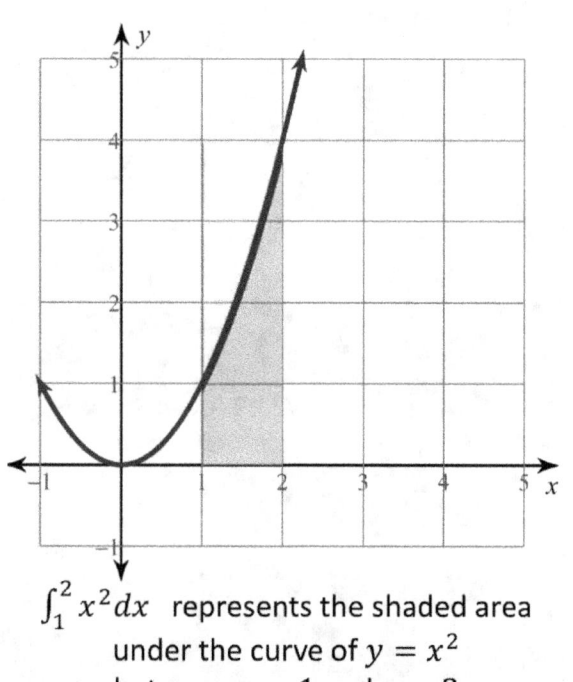

$\int_1^2 x^2 dx$ represents the shaded area under the curve of $y = x^2$ between $x = 1$ and $x = 2$

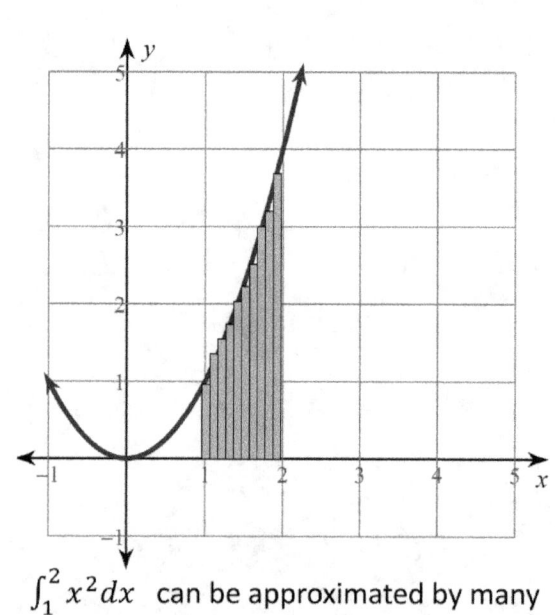

$\int_1^2 x^2 dx$ can be approximated by many small rectangles

Area above the x-axis is considered positive while area below the x-axis is considered negative. With very simple functions, we can just use geometry to break up a complicated shape to find the area under the curve

Ex. 1

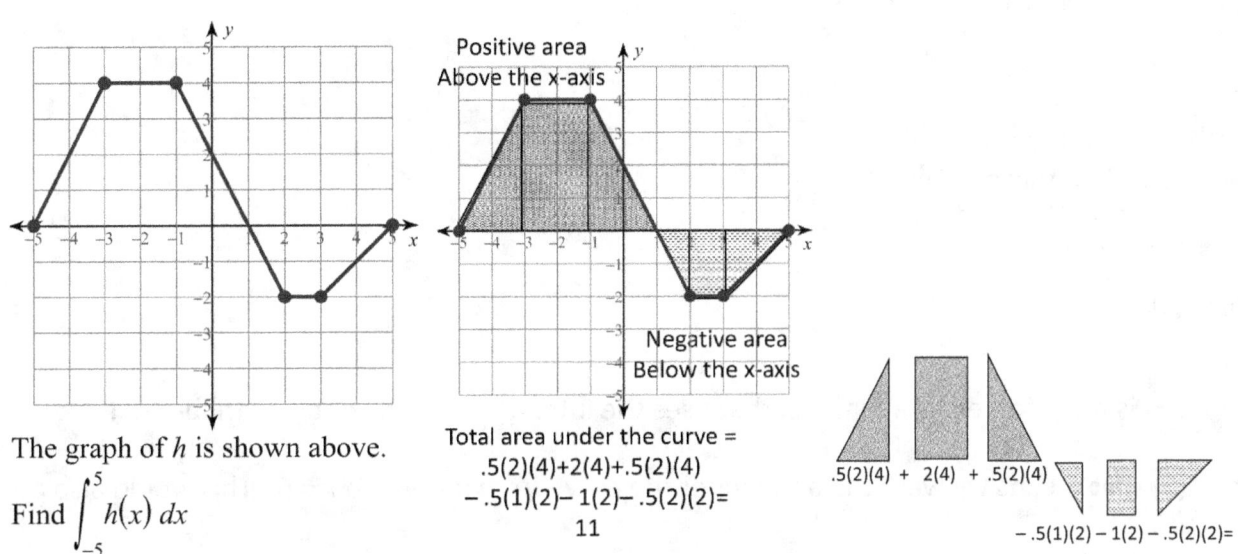

The graph of h is shown above.

Find $\int_{-5}^{5} h(x)\, dx$

Total area under the curve =
.5(2)(4)+2(4)+.5(2)(4)
− .5(1)(2)− 1(2)− .5(2)(2)=
11

.5(2)(4) + 2(4) + .5(2)(4)
− .5(1)(2) − 1(2) − .5(2)(2)=

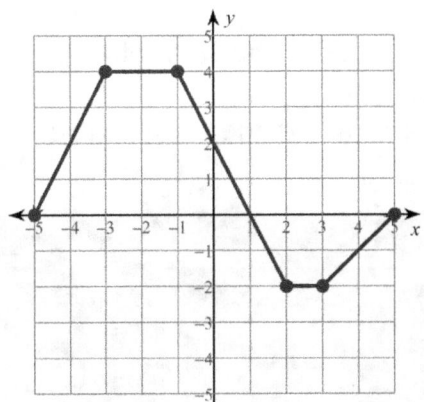

The graph of h is shown above.

Therefore $\int_{-5}^{5} h(x)\, dx = 11$

Another way this problem could be stated:

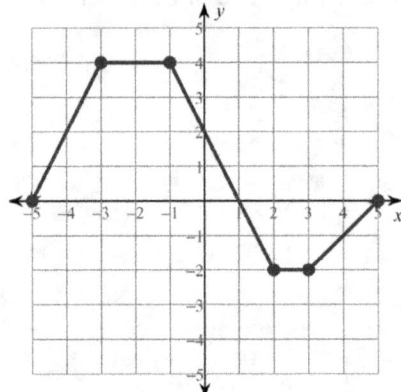

The graph of *h* is shown above.

Let $g(x) = \int_{-5}^{x} h(t)\, dt$

Find $g(5)$

Now g(x) is represented as a function that uses the integration from – 5 to a variable x. If we find g(5) that means we want the area under the curve from x = – 5 to x = 5. This would also be evaluated to be 11.

When doing these types of integrals with simple geometric shapes, we find the areas of the triangles, rectangles, and trapezoids. Again, remember that any area above the x-axis will be counted as positive while areas below the x-axis is negative.

Practice 33

1)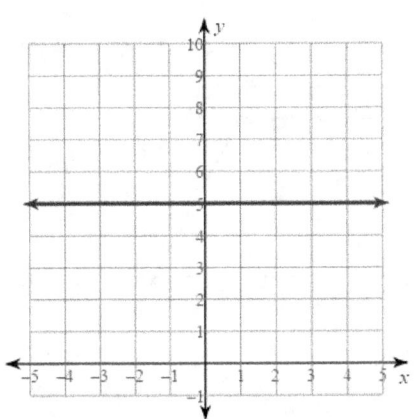

The graph of h is shown above.

Find $\displaystyle\int_{-2}^{4} h(x)\,dx$

2)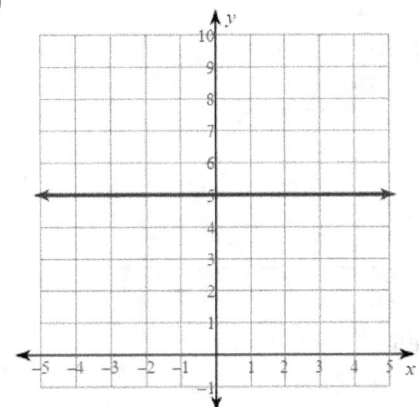

The graph of h is shown above.

Let $g(x) = \displaystyle\int_{-2}^{x} h(t)\,dt$

Find $g(2)$

3)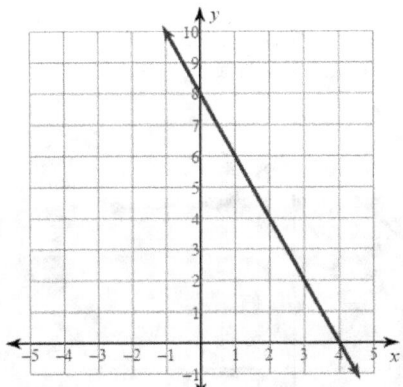

The graph of h is shown above.

Find $\displaystyle\int_{0}^{4} h(x)\,dx$

4)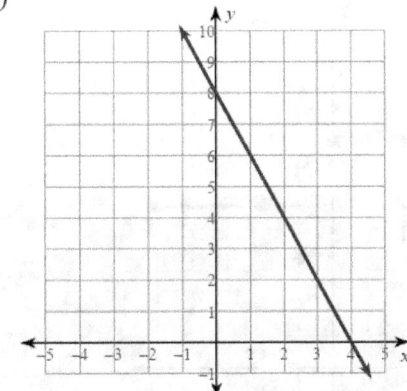

The graph of h is shown above.

Let $g(x) = \displaystyle\int_{0}^{x} h(t)\,dt$

Find $g(2)$

5)

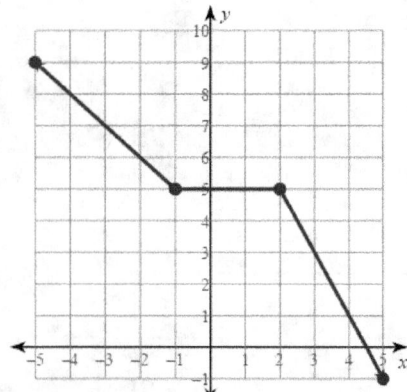

The graph of h is shown above.

Let $g(x) = \int_{-2}^{x} h(t)\, dt$

Find $g(3)$

6)

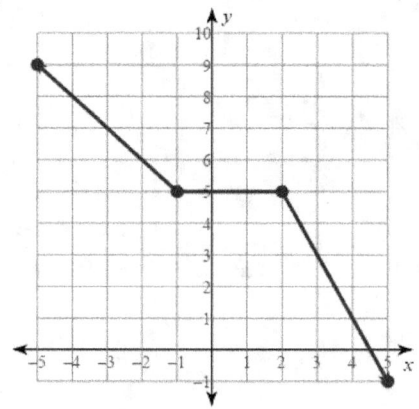

The graph of h is shown above.

Let $g(x) = \int_{-4}^{x} h(t)\, dt$

Find $g(4)$

7)

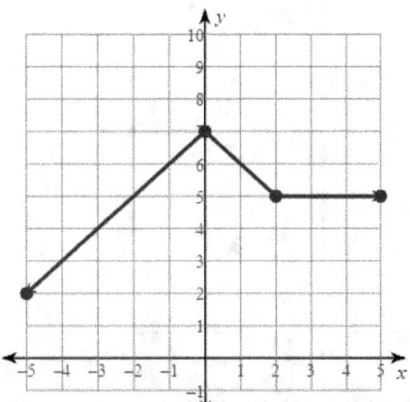

The graph of h is shown above.

Let $g(x) = \int_{-2}^{x} h(t)\, dt$

Find $g(3)$

8)

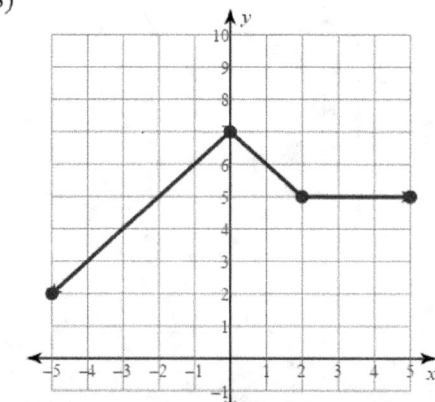

The graph of h is shown above.

Let $g(x) = \int_{-3}^{x} h(t)\, dt$

Find $g(4)$

9)

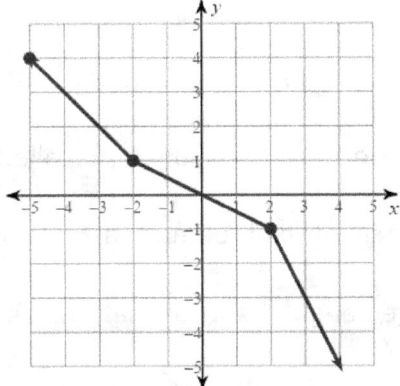

The graph of h is shown above.

Let $g(x) = \int_{-4}^{x} h(t)\, dt$

Find $g(4)$

10)

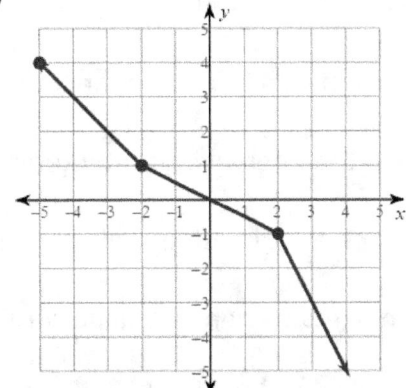

The graph of h is shown above.

Let $g(x) = \int_{-2}^{x} h(t)\, dt$

Find $g(3)$

11)

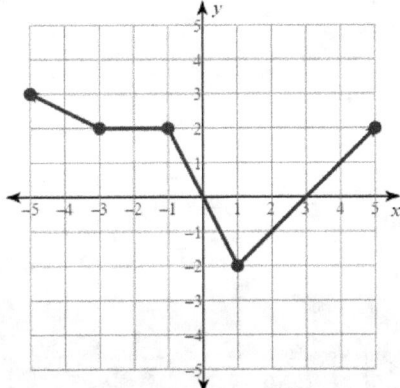

The graph of h is shown above.

Let $g(x) = \int_{-3}^{x} h(t)\, dt$

Find $g(3)$

12)

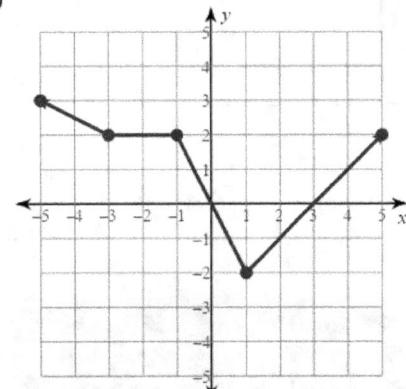

The graph of h is shown above.

Let $g(x) = \int_{-5}^{x} h(t)\, dt$

Find $g(5)$

Approximating Area with Riemann Sums

A Riemann sum is a way of approximating the area under a curve using rectangles. We usually split up the interval into subintervals of equal widths. A left Riemann sum, also called a left rectangular approximation, will use the left endpoints of each interval to find the appropriate heights. A right Riemann sum, also called a right rectangular approximation, will use the right endpoints for the heights. A midpoint Riemann sum, also called a midpoint rectangular approximation, will use the heights at the midpoints of each subinterval. A trapezoidal approximation will use trapezoids constructed from the subintervals.

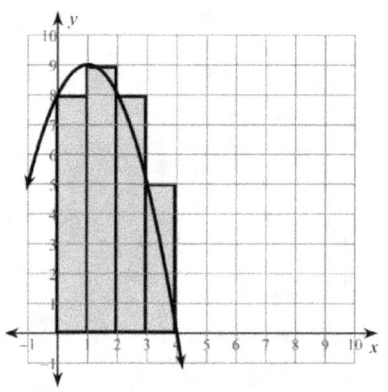
Left Riemann Sum
Left Rectangular Approximation Method
LRAM

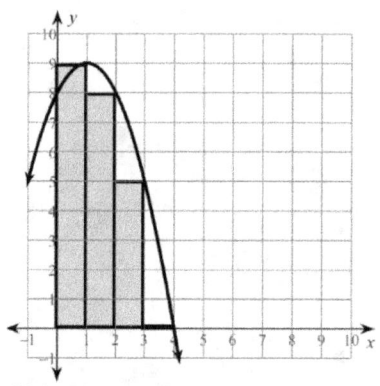
Right Riemann Sum
Right Rectangular Approximation Method
RRAM

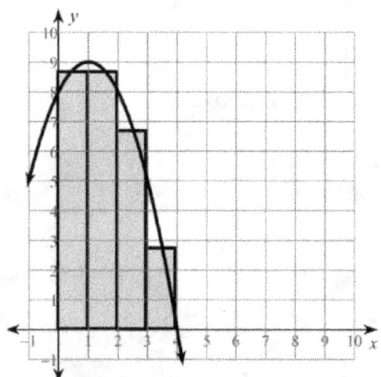
Midpoint Riemann Sum
Midpoint Rectangular Approximation Method
MRAM

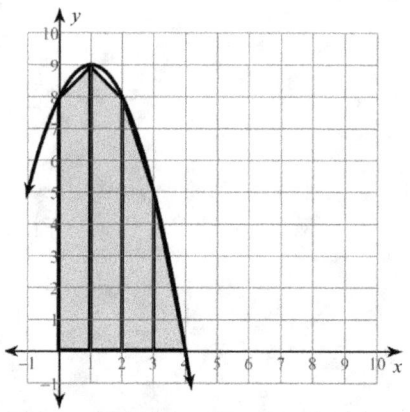
Trapezoidal Approximation Method
TRAP

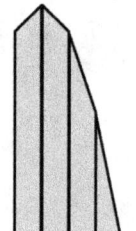
Trapezoidal Approximation Method
TRAP

These four methods are the main methods of approximating areas under a curve. As you can see from the examples above, sometimes you get an overestimate, sometimes an underestimate. One type of approximation may be better for a certain function, while another approximation is better for a different function. We will be exploring more of this later.

Ex. 1 For each problem, use a left-hand Riemann sum to approximate the integral based off of the values in the table. Then use a right-hand Riemann sum to approximate the integral.

$$\int_0^{10} f(x)\, dx$$

x	0	1	5	7	10
$f(x)$	2	4	6	4	5

Solution: Plot the points on a graph and draw rectangles using heights from the left endpoints. Calculate the areas of all the rectangles formed and add up the areas. Notice that LRAM will use the $f(x)$ values starting on the left, using 2, 4, 6, 4. $A = 1(2) + 4(4) + 2(6) + 3(4) = 42$. *By LRAM, the approximation is 42 square units.* Now we will calculate by RRAM. Plot the points on a graph and draw rectangles using heights from the right endpoints. Calculate the areas of all the rectangles formed and add up the areas. Notice that RRAM will use the $f(x)$ values starting on the right, using 4, 6, 4, 5. $A = 1(4) + 4(6) + 2(4) + 3(5) = 51$. *By LRAM, the approximation is 51 square units.* These are just estimates, the actual area should be close to these. After practicing and seeing the pattern, it is not always necessary to draw the picture.

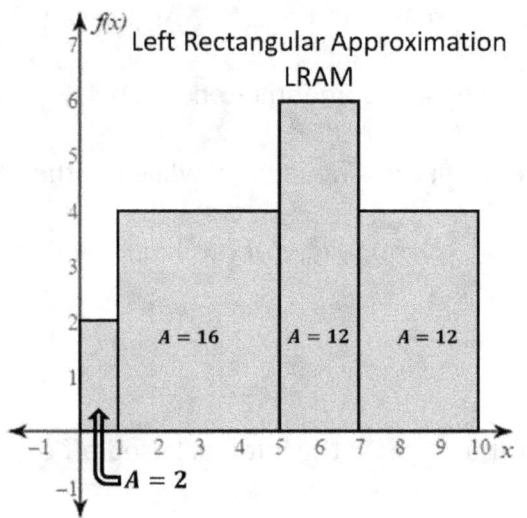

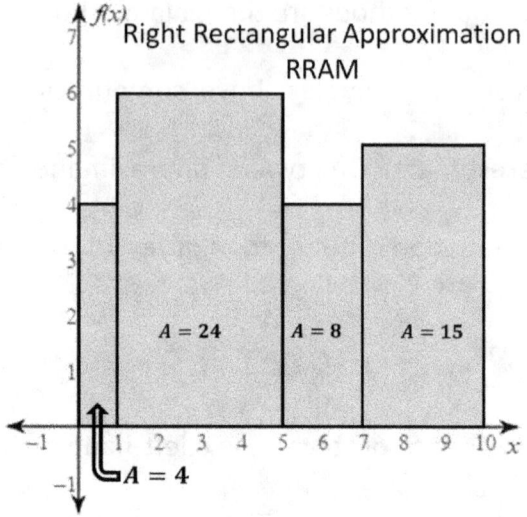

Practice 34

For each problem, use a left-hand Riemann sum to approximate the integral based off of the values in the table.

1) $\int_0^8 f(x)\,dx$

x	0	1	6	7	8
f(x)	2	3	5	4	2

2) $\int_0^8 f(x)\,dx$

x	0	1	2	7	8
f(x)	6	5	4	5	3

3) $\int_0^8 f(x)\,dx$

x	0	1	2	6	8
f(x)	7	5	7	6	4

4) $\int_0^8 f(x)\,dx$

x	0	2	4	6	8
f(x)	7	6	5	6	5

5) $\int_0^9 f(x)\,dx$

x	0	3	4	5	9
f(x)	0	1	0	−1	−2

6) $\int_0^9 f(x)\,dx$

x	0	1	5	8	9
f(x)	−1	−2	−3	−4	−3

For each problem, use a right-hand Riemann sum to approximate the integral based off of the values in the table.

7) $\int_0^{10} f(x)\,dx$

x	0	2	6	9	10
f(x)	7	8	9	7	5

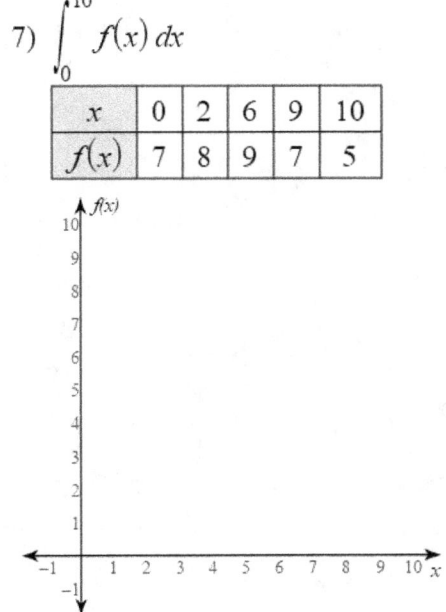

8) $\int_0^8 f(x)\,dx$

x	0	1	3	5	8
f(x)	4	6	7	5	6

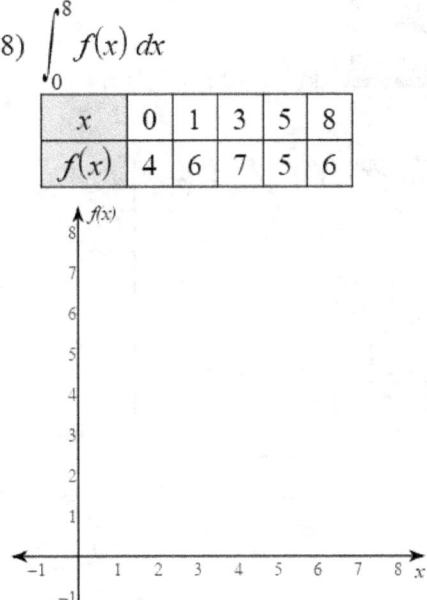

9) $\int_0^9 f(x)\,dx$

x	0	2	6	7	9
f(x)	7	6	5	4	2

10) $\int_0^9 f(x)\,dx$

x	0	1	2	4	9
f(x)	2	4	6	5	4

11) $\int_0^8 f(x)\,dx$

x	0	2	4	5	8
f(x)	-3	-4	-3	-2	-3

12) $\int_0^{10} f(x)\,dx$

x	0	1	3	6	10
f(x)	-3	-2	-1	0	1

GRAPHING CALCULATOR: DEFINITE INTEGRAL

In this section we will go over the steps to find a definite integral using the calculator.

Press the 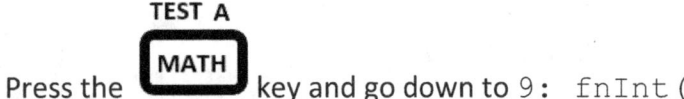 key and go down to 9: fnInt(

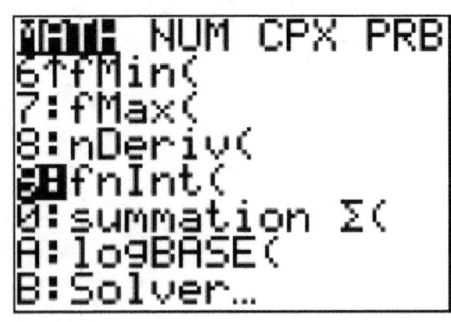

The screen should look like this and you can type in your variable, function, lower and upper bounds of your definite integral. For example, if we want the derivative of $\int_1^3 x^2 \, dx$ we type in everything and press [ENTER]. We get an answer of 8. So $f'(4) = 8$.

If you have an older calculator your screen may look like this:

Approximating Area with Trapezoidal Rule

Trapezoidal Approximation Formula: To approximate the area under a curve between *a* and *b* using the trapezoidal approximation with *n* subintervals:

$$T_n = \frac{b-a}{2n}[(f(x_0) + 2f(x_1) + 2f(x_2) + \cdots + 2f(x_{n-1}) + f(x_n)]$$

Ex 2. For each problem approximate the area with 4 subdivisions using Left Rectangular Approximation, Left Rectangular Approximation, Midpoint Rectangular Approximation, Trapezoidal Approximation, and Actual Area. You may use a calculator.

$f(x) = -\dfrac{x^2}{2} + 3x;\quad [0, 4]$

LRAM:

LRAM:

MRAM:

TRAP:

ACTUAL:

Solution: To do this type of problem with so many parts and calculations, a calculator will definitely be handy. Type in the function in y =. Then adjust the table to get the endpoints we need. Since we are going from 0 to 4 with 4 intervals, each interval will be 1 unit in width.

For the LRAM: We will have 1(0) + 1(2.5) + 1(4) + 1(4.5) = 11

For RRAM: 1(2.5) + 1(4) + 1(4.5) + 1(4) = 15

For the midpoint, adjust the table to get the midpoint values

For MRAM: 1(1.375) + 1(3.375) + 1(4.375) + 1(4.375) = 13.5

For Trapezoidal Rule (TRAP) we use the formula: .5[(0) + 2(2.5) + 2(4) + 2(4.5)+4] = *13*

The actual area can be found using the calculator using the MATH button. Actual is *13.3333*

$$\int_0^4 \left(-\frac{x^2}{2}+3x\right)dx = 13.33333333$$

Practice 35

For each problem approximate the area with 4 subdivisions using each method and find actual.

1) $f(x) = -\dfrac{x^2}{2} + x + 5$; [0, 4]

   LRAM:

   LRAM:

   MRAM:

   TRAP:

   ACTUAL:

2) $f(x) = \dfrac{x^2}{2} + x + 2$; [−5, −1]

   LRAM:

   LRAM:

   MRAM:

   TRAP:

   ACTUAL:

3) $f(x) = \dfrac{5}{x}$; [1, 5]

   LRAM:

   LRAM:

   MRAM:

   TRAP:

   ACTUAL:

4) $f(x) = -\dfrac{4}{x}$; [−6, −2]

   LRAM:

   LRAM:

   MRAM:

   TRAP:

   ACTUAL:

Order the approximations from smallest to largest: Right Rectangular Approximation, Left Rectangular Approximation, Trapezoidal Approximation, Actual Area

5)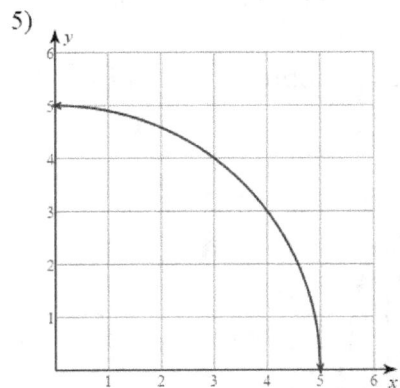

6)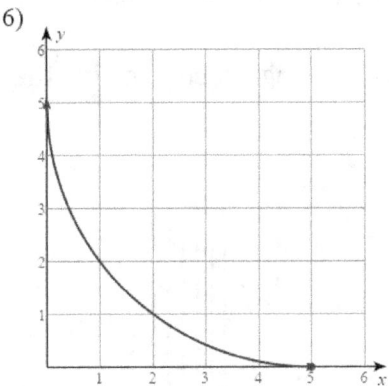

7)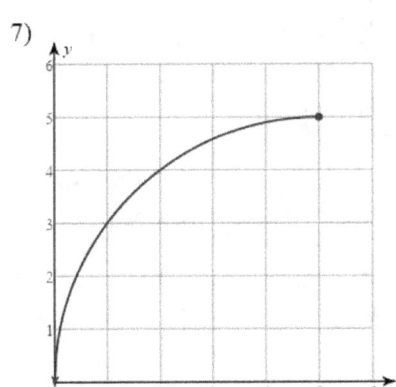

8)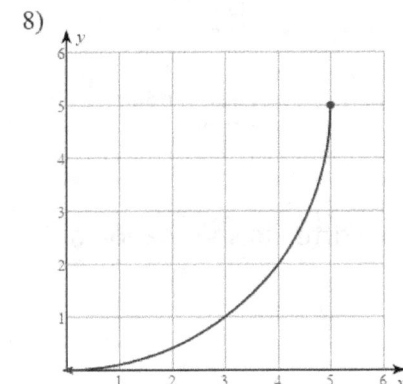

Order the areas Left Riemann Sum, Right Riemann Sum, Actual Area Under the Curve from least to greatest.

9) Function $g(x)$ is continuous and decreasing. We want to find the area under the curve between $x = 0$ and $x = 12$.

10) Function $g(x)$ is continuous and increasing. We want to find the area under the curve between $x = 0$ and $x = 12$.

Riemann Sums and Definite Integrals

Riemann Sums are the process of finding the area under the curve using rectangles. The more intervals or rectangles we have the better our approximation will be. One way to define a definite integral is by a limiting process letting the number of intervals approach infinity.

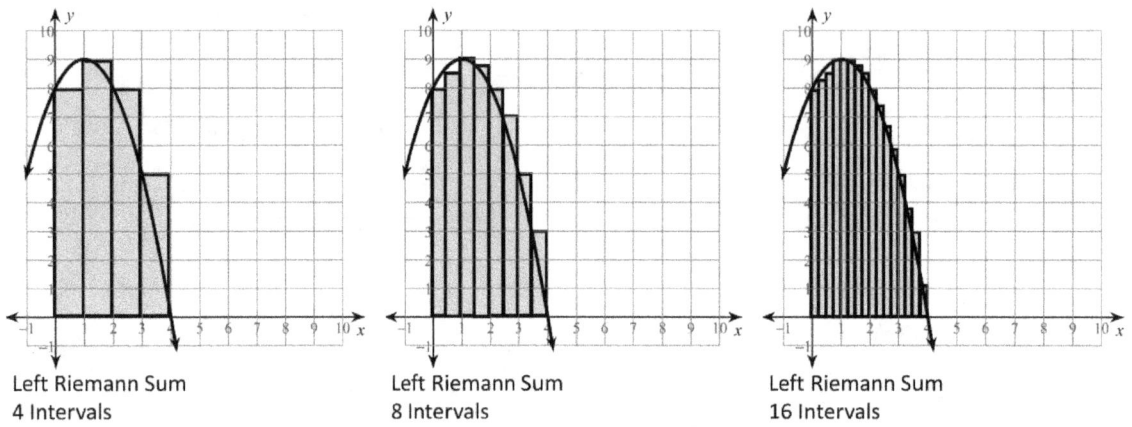

Left Riemann Sum
4 Intervals

Left Riemann Sum
8 Intervals

Left Riemann Sum
16 Intervals

A Riemann Sum can be written in summation notation:

$$\sum_{i=m}^{n} f(x_i) \Delta x$$

Where $f(x_i)$ is the height of the rectangle at the ith position and Δx is the width of each interval. We can have a left Riemann sum or a Right Riemann sum depending on where we start counting the intervals. These are defined below:

$$\sum_{i=0}^{n-1} f(x_i) \cdot \Delta x \qquad \sum_{i=1}^{n} f(x_i) \cdot \Delta x$$

LEFT RIEMANN SUM   RIGHT RIEMANN SUM

$$\Delta x = \frac{b-a}{n}$$

Width of the interval
Length/number of intervals

$$x_i = a + \Delta x \cdot i$$

Current position is starting
position plus each interval

Ex. 1 Write a Left Riemann Sum that approximates the area between $f(x) = \frac{20}{x+2}$ and the x-axis on the interval [1,4] with 6 equal subdivisions?

Solution: Note that in this example we have the function given, also from the interval, $a = 1$ and $b = 4$. First we find the width of the intervals $\Delta x = \frac{4-1}{6} = .5$ so the width is .5. Next, we express $x_i = 1 + .5i$. This means that $f(x_i) = \frac{20}{1+.5i+2} = \frac{20}{3+.5i}$. Now we can put this into the sigma notation. Since this is a left Riemann sum we start our counter at $i = 0$: $\sum_{i=0}^{n-1}(\frac{20}{3+.5i} \cdot .5)$ or $\sum_{i=0}^{n-1}(\frac{10}{3+.5i})$

A definite integral can be defined in terms of the limit of a Riemann Sum. It doesn't matter if we use a left hand, right hand or any other commonly defined method; as we get infinitely many slices of infinitesimal size, they all approach the same limit.

$$\int_a^b f(x)dx = \lim_{n \to \infty} \sum_{i=1}^{n} f(x_i) \cdot \Delta x$$

Where $\Delta x = \frac{b-a}{n}$ and $x_i = a + \Delta x \cdot i$

Ex. 2 Write a limit of a Riemann Sum that is equivalent to $\int_0^\pi \cos x \, dx$

Solution: Note that $a = 0$ and $b = \pi$. First, we find the width $\Delta x = \frac{\pi - 0}{n} = \frac{\pi}{n}$. Next find $x_i = 0 + \frac{\pi}{n}i$. Now we write $\lim_{n \to \infty} \sum_{i=1}^{n} \cos(\frac{\pi}{n}i) \cdot \frac{\pi}{n}$

Therefore, a possible answer is: $\lim_{n \to \infty} \sum_{i=1}^{n} \cos(\frac{\pi i}{n}) \cdot \frac{\pi}{n}$

Practice 36

1) Write a Left Riemann Sum to approximate the area between $f(x) = -|x| + 4$ and the x-axis on the interval $[-4, 4]$ with 8 equal subdivisions?

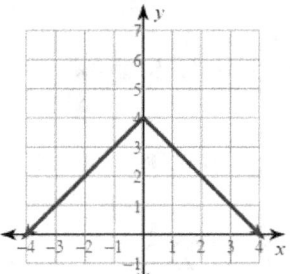

2) Write a Right Riemann Sum to approximate the area between $f(x) = -|x| + 4$ and the x-axis on the interval $[-4, 4]$ with 8 equal subdivisions?

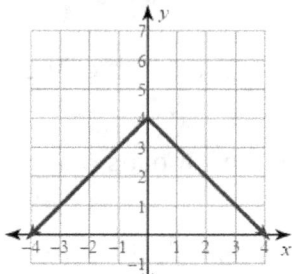

3) Write a Right Riemann Sum to approximate the area between $f(x) = -|x| + 4$ and the x-axis on the interval $[-4, 4]$ with 4 equal subdivisions?

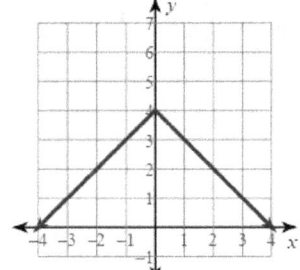

4) Write a Left Riemann Sum to approximate the area between $f(x) = -|x| + 4$ and the x-axis on the interval $[-4, 4]$ with 4 equal subdivisions?

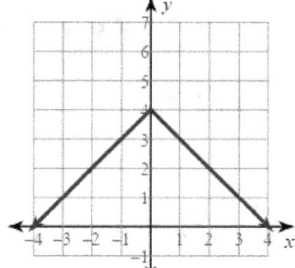

5) Write a Left Riemann Sum to approximate the area between $f(x) = -|x| + 4$ and the x-axis on the interval $[-4, 4]$ with 16 equal subdivisions?

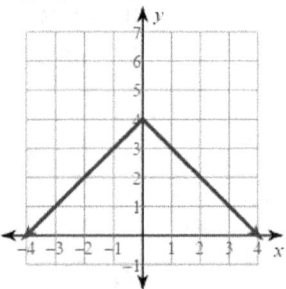

6) Write a Right Riemann Sum to approximate the area between $f(x) = -|x| + 4$ and the x-axis on the interval $[-4, 4]$ with 16 equal subdivisions?

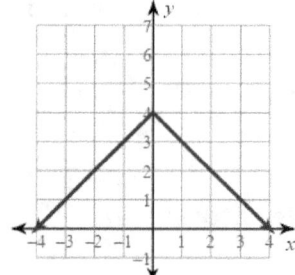

7) Which of the following approximates the area between $f(x) = -|x| + 6$ and the x-axis on the interval $[-1, 3]$ using a right Riemann sum with 4 equal subdivisions?

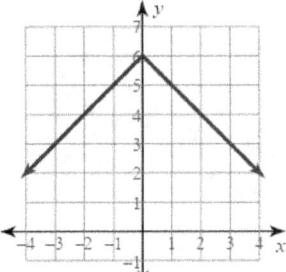

8) Which of the following approximates the area between $f(x) = -x^2 + 4$ and the x-axis on the interval $[-2, 2]$ using a right Riemann sum with 4 equal subdivisions?

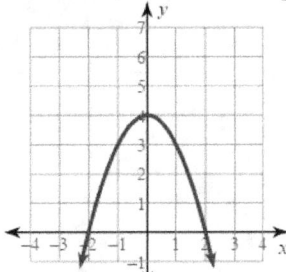

9) Which of the following approximates the area between $f(x) = -(x - 1)^2 + 4$ and the x-axis on the interval $[-1, 3]$ using a left Riemann sum with 4 equal subdivisions?

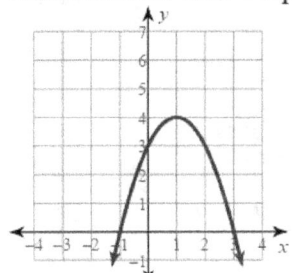

10) Which of the following approximates the area between $f(x) = -|x| + 6$ and the x-axis on the interval $[-1, 3]$ using a left Riemann sum with 4 equal subdivisions?

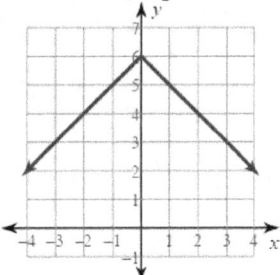

11) Which of the following approximates the area between $f(x) = -|x + 2| + 5$ and the x-axis on the interval $[-3, 3]$ using a right Riemann sum with 12 equal subdivisions?

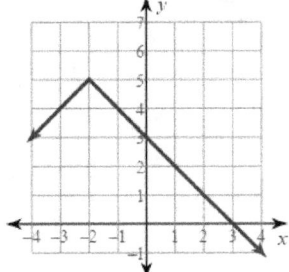

12) Which of the following approximates the area between $f(x) = -|x - 1| + 5$ and the x-axis on the interval $[-3, 1]$ using a left Riemann sum with 16 equal subdivisions?

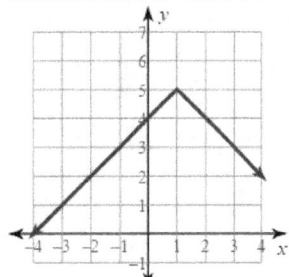

13) Rewrite as a limit
$$\int_0^\pi \cos x \, dx$$

14) Rewrite as a limit
$$\int_0^\pi \sin x \, dx$$

15) Rewrite as a limit
$$\int_1^e \ln x \, dx$$

16) Rewrite as a limit
$$\int_{\frac{\pi}{2}}^\pi \sin x \, dx$$

17) Rewrite as an integral
$$\lim_{n \to \infty} \sum_{i=1}^{n} \left(4 + \frac{4i}{n}\right)^2 \cdot \frac{4}{n}$$

18) Rewrite as an integral
$$\lim_{n \to \infty} \sum_{i=0}^{n-1} \sqrt{9 + \frac{7i}{n}} \cdot \frac{7}{n}$$

19) Rewrite as an integral
$$\lim_{n \to \infty} \sum_{i=0}^{n-1} \ln\left(1 + \frac{ie - i}{n}\right) \cdot \frac{e - 1}{n}$$

20) Rewrite as an integral
$$\lim_{n \to \infty} \sum_{i=1}^{n} e^{1 + \frac{4i}{n}} \cdot \frac{4}{n}$$

Properties of Definite Integrals

Properties for Definite Integrals

$\int_a^a f(x)dx = 0$ The integral at a single point is zero. There is no area for a rectangle of width zero.

$\int_a^b f(x)dx = -\int_b^a f(x)dx$ The accumulation from a to b is the opposite of going backwards from b to a.

$\int_a^b k \cdot f(x)dx = k\int_a^b f(x)dx$ A constant can be factored in/out of an integral.

$\int_a^b (f(x) + g(x))dx = \int_a^b f(x)dx + \int_a^b g(x)dx$ Add integrals

$\int_a^b (f(x) - g(x))dx = \int_a^b f(x)dx - \int_a^b g(x)dx$ Subtract integrals

$\int_a^c f(x)dx + \int_c^b f(x)dx = \int_a^b f(x)dx$ We can add up adjacent integrals to form one integral

Practice 37

1) If $\int_2^8 f(x)\,dx = -2$ and $\int_2^8 g(x)\,dx = 6$,
Find

$\int_8^2 g(x)\,dx$

$\int_2^8 (f(x) + g(x))\,dx$

$\int_8^2 3g(x)\,dx$

$\int_8^8 f(x)\,dx$

$\int_2^8 (f(x) - g(x))\,dx$

$\int_2^4 f(x)\,dx + \int_4^8 f(x)\,dx$

$\int_8^2 (2f(x) - 3g(x))\,dx$

2) If $\int_a^b f(x)\,dx = 2x + 4$; $\int_a^b g(x)\,dx = 3x - 6$,
Find

$\int_a^b (f(x) + g(x))\,dx$

$\int_b^a g(x)\,dx$

$\int_b^a 3g(x)\,dx$

$\int_a^a f(x)\,dx$

$\int_b^a (f(x) - g(x))\,dx$

$\int_b^a (3f(x) - 2g(x))\,dx$

$\int_a^c 2f(x)\,dx + \int_c^b 2f(x)\,dx$

3) $\int_{-2}^{6} f(x)\,dx = 2$

$\int_{-2}^{6} g(x)\,dx = 6$

$\int_{6}^{10} f(x)\,dx = 14$

$\int_{6}^{10} g(x)\,dx = -8$

Find:

$\int_{-2}^{10} f(x)\,dx$

$\int_{-2}^{10} g(x)\,dx$

$\int_{-2}^{-2} f(x)\,dx$

$\int_{-2}^{6} (f(x) + g(x))\,dx$

$\int_{6}^{10} 2f(x)\,dx$

$\int_{2}^{2} f(x)\,dx$

$\int_{10}^{6} f(x)\,dx$

$\int_{-2}^{10} (f(x) + g(x))\,dx$

$\int_{6}^{-2} (f(x) - g(x))\,dx$

$\int_{10}^{-2} (3f(x) - 2g(x))\,dx$

4) $\int_{a}^{b} f(x)\,dx = 2x + 1$

$\int_{a}^{b} g(x)\,dx = 3x - 2$

$\int_{b}^{c} f(x)\,dx = -x - 4$

$\int_{b}^{c} g(x)\,dx = -4x + 6$

Find:

$\int_{a}^{c} g(x)\,dx$

$\int_{a}^{c} f(x)\,dx$

$\int_{b}^{c} 3f(x)\,dx$

$\int_{b}^{b} f(x)\,dx$

$\int_{a}^{b} (f(x) + g(x))\,dx$

$\int_{b}^{b} (f(x) + g(x))\,dx$

$\int_{c}^{b} f(x)\,dx$

$\int_{b}^{a} (f(x) - g(x))\,dx$

$\int_{a}^{c} (f(x) + g(x))\,dx$

$\int_{c}^{a} (2f(x) - 3g(x))\,dx$

Fundamental Theorem of Calculus

It has been implied and you may have guessed that there is a relationship between differentiation and integration. Differentiation and integration are essentially inverse operations. Sometimes integration is also called antidifferentiation. The fundamental theorem of calculus makes this relationship more explicit.

Fundamental Theorem of Calculus: Suppose that the function $f(x)$ is continuous on an interval containing the point a. Then

Part I: $\frac{dy}{dx} \cdot \int_a^x f(t)dt = f(x)$ Basically the derivative undoes the integral

Part II: $\int_a^b f(x)dx = F(b) - F(a)$ Where $F(x)$ is the antiderivative of $f(x)$,

i.e. $F'(x) = f(x)$

<u>Ex. 1</u> $F(x) = \int_{-4}^x (t^2 - 3t + 6)\, dt$ Find $F'(x)$

Solution: We can use the first part of the fundamental theorem of calculus. We are taking the derivative of an integral which will undo the function. The lower bound, in this case -4, does not affect the derivative (the derivative of a constant is zero). *Therefore, $F'(x) = x^2 - 3x + 6$.*

<u>Ex. 2</u> $g(x) = \int_{-4}^x \sqrt{t+4}\, dt$ Find $g'(12)$

Solution: Again, we can use the first part of the fundamental theorem of calculus. This time we are going to evaluate the function g' at 12. *Therefore, $g'(12) = \sqrt{12+4} = 4$.*

Ex. 3 $F(x) = \int_{-4}^{x^2}(-4t + 3)\,dt$ Find $F'(x)$

Solution: Here's our last example using the first part of the fundamental theorem of calculus. This one is a little more complicated because the upper bound is not just x but a function of x.

Usually if $F(x) = \int_a^x h'(t)dt$, then $F'(x) = h'(x)$, but here we have $F(x) = \int_a^{g(x)} h'(t)dt$, so that means by the chain rule we would have $F'(x) = h'(g(x))g'(x)$. Okay here we have

$F(x) = \int_{-4}^{x^2}(-4t + 3)\,dt$, that means that we plug in the x^2 for x and use the chain rule:

$F'(x) = (-4(x^2) + 3) \cdot \frac{dy}{dx}x^2$ This then gives: $(-4x^2 + 3)(2x) = -8x^3 + 6x$. Therefore,

$F'(x) = -8x^3 + 6x$

Practice 38

1) $F(x) = \int_{-3}^{x} (t^2 + 4t - 1)\,dt$
   Find $F'(x)$

2) $F(x) = \int_{-7}^{x} (-t^2 - 8t - 13)\,dt$
   Find $F'(x)$

3) $F(x) = \int_{-5}^{x} (t^2 + 6t + 11)\,dt$
   Find $F'(x)$

4) $F(x) = \int_{1}^{x} (2t - 1)\,dt$
   Find $F'(x)$

5) $g(x) = \displaystyle\int_{2}^{x} \sqrt{t}\, dt$

Find $g'(16)$

6) $g(x) = \displaystyle\int_{-\frac{\pi}{2}}^{x} \sin t\, dt$

Find $g'\left(\dfrac{\pi}{2}\right)$

7) $g(x) = \displaystyle\int_{2}^{x} \sqrt[3]{t+2}\, dt$

Find $g'(25)$

8) $g(x) = \displaystyle\int_{-6}^{x} (t+2)^2\, dt$

Find $g'(4)$

9) $g(x) = \displaystyle\int_{-2}^{x} (t+2)^3\, dt$

Find $g'(2)$

10) $g(x) = \displaystyle\int_{1}^{x} \dfrac{1}{t}\, dt$

Find $g'(2)$

11) $F(x) = \displaystyle\int_{-\frac{\pi}{4}}^{x} -\sin t\, dt$

Find $F'(x)$

12) $F(x) = \displaystyle\int_{-\frac{\pi}{6}}^{x} 2\sec^2 t\, dt$

Find $F'(x)$

13) $F(x) = \int_0^{x^2} (-2t + 2)\, dt$

Find $F'(x)$

14) $F(x) = \int_{-2}^{x^3} (-2t + 1)\, dt$

Find $F'(x)$

15) $F(x) = \int_{-3}^{x^2} (t^2 + 2t)\, dt$

Find $F'(x)$

16) $F(x) = \int_{-4}^{2x} (2t + 2)\, dt$

Find $F'(x)$

17) $F(x) = \int_0^{x^3} (-t^3 + 3t^2 - 2)\, dt$

Find $F'(x)$

18) $F(x) = \int_0^{x^3} (t + 1)\, dt$

Find $F'(x)$

19) $F(x) = \int_{-\frac{\pi}{2}}^{x^2} -\sin t\, dt$

Find $F'(x)$

20) $F(x) = \int_{\frac{\pi}{4}}^{x^3} \csc t \cot t\, dt$

Find $F'(x)$

Finding Antiderivatives and Indefinite Integrals

When we take the derivative of x^2 we get 2x. That makes the antiderivative of 2x equal to x^2. Notice that x^2 is not the only function that has a derivative of 2x. There are infinitely many, $x^2 + 1$, $x^2 + 2$, $x^2 + 3$,... Any constant added to the end will also give the same derivative. Therefore, we write all the solutions to the antiderivative of 2x as $x^2 + C$, where C is any constant.

To determine what the antiderivative is for a function, we just have to think about what function we can differentiate to get it. For some functions, this may be relatively easy, but for other types it may be more difficult, and there are even some functions that we cannot integrate in a finite combination of elementary functions.

Reverse Power Rule

Reverse Power Rule for Integrals: $\int u^n du = \frac{u^{n+1}}{n+1} + C$, when $n \neq -1$, C is a constant

Basically the power rule states that we bump up the exponent by one, and then divide by that number. The power rule doesn't work for n = −1 because we would end up dividing by zero which is undefined: $\int x^{-1} dx = \frac{x^{-1+1}}{-1+1} = \frac{x^0}{0} = undefined$

<u>Ex. 1</u> Solve $\int (x^2 + 2x - 4)\, dx$

<u>Solution:</u> Properties of integrals allows us to separate this into: $\int x^2 dx + \int 2x\, dx - \int 4\, dx$.

We can use the power rule on each term: $\frac{x^3}{3} + \frac{2x^2}{2} - \frac{4x}{1} + C = \frac{x^3}{3} + x^2 - 4x + C$. We can always check by taking the derivative. Notice that $\frac{dy}{dx}\left(\frac{x^3}{3} + x^2 - 4x + C\right) = \frac{3x^2}{3} + 2x - 4 + 0 = x^2 + 2x - 4$. Therefore, $\int (x^2 + 2x - 4)\, dx = \frac{x^3}{3} + x^2 - 4x + C$

The strength of the power rule is that not only polynomial functions like above, but rational and radical functions can usually be written in terms of exponents, allowing the power rule to be applied.

Ex. 2 Solve $\int \frac{1}{x^2}\, dx$

Solution: This function can be rewritten in terms of powers as: $\int \frac{1}{x^2}\, dx = \int x^{-2}\, dx$. We can again use the power rule: $\frac{x^{-1}}{-1} + C = -\frac{1}{x} + C$

Ex. 3 Solve $\int \sqrt[3]{x} + \sqrt{x}\, dx$

Solution: Here we need to rewrite radicals in rational exponent form as: $\int \sqrt[3]{x} + \sqrt{x}\, dx = \int x^{\frac{1}{3}} + x^{\frac{1}{2}}\, dx$. We can again use the power rule: $\frac{x^{\frac{4}{3}}}{\frac{4}{3}} + \frac{x^{\frac{3}{2}}}{\frac{3}{2}} + C = \frac{3x^{4/3}}{4} + \frac{2x^{3/2}}{3} + C = \frac{3}{4}\sqrt[3]{x^4} + \frac{2}{3}\sqrt{x^3} + C$

Practice 39

1) $\int (10x^4 - 8x^3 + 1)\, dx$

2) $\int (15x^4 + 9x^2 + 8x)\, dx$

3) $\int (30x^5 + 15x^2 + 8x)\, dx$

4) $\int (-12x^5 + 15x^4 - 1)\, dx$

5) $\int (8x^3 - 10x + 3)\, dx$

6) $\int (25x^4 - 3x^2 - 4)\, dx$

7) $\int \left(\dfrac{3}{x^2} + \dfrac{12}{x^4} - \dfrac{4}{x^5} \right) dx$

8) $\int \left(\dfrac{6}{x^3} - \dfrac{9}{x^4} - \dfrac{4}{x^5} \right) dx$

9) $\int (4x^{-3} - 9x^{-4} + 12x^{-5})\, dx$

10) $\int \left(-\dfrac{4}{x^2} - \dfrac{12}{x^4} + \dfrac{20}{x^5}\right) dx$

11) $\int \left(-\dfrac{1}{x^2} - \dfrac{6}{x^3} - \dfrac{12}{x^5}\right) dx$

12) $\int \left(-\dfrac{2}{x^2} - \dfrac{2}{x^3} - \dfrac{15}{x^4}\right) dx$

13) $\int \left(\dfrac{5x^{\frac{3}{2}}}{2} + \dfrac{36x^{\frac{5}{4}}}{4} + \dfrac{12x^{\frac{1}{3}}}{3}\right) dx$

14) $\int \left(\dfrac{10\sqrt[3]{x^2}}{3} + \dfrac{8\sqrt[3]{x}}{3} + 5\sqrt[4]{x}\right) dx$

15) $\int \left(\dfrac{10\sqrt[3]{x^2}}{3} + 7\sqrt[5]{x^2} - 6\sqrt[5]{x}\right) dx$

16) $\int \left(\dfrac{25\sqrt[3]{x^2}}{3} - \dfrac{14\sqrt[5]{x^2}}{5} + 6\sqrt[5]{x}\right) dx$

17) $\displaystyle\int\left(\dfrac{16x^{\frac{5}{3}}}{3}+\dfrac{7x^{\frac{4}{3}}}{3}+\dfrac{6x^{\frac{1}{5}}}{5}\right)dx$

18) $\displaystyle\int\left(\dfrac{14\sqrt[5]{x^2}}{5}+\dfrac{25\sqrt[4]{x}}{4}+\dfrac{6\sqrt[5]{x}}{5}\right)dx$

Simplify first, then find the integral.

19) $\displaystyle\int 3x(4x^4+5x-2)\,dx$

20) $\displaystyle\int x^2(15x^2+16x+12)\,dx$

21) $\displaystyle\int (x+1)(8x^2-5x+5)\,dx$

22) $\displaystyle\int 6x(3x^2-1)(x^2+1)\,dx$

23) $\displaystyle\int \dfrac{-x^3-15x-4}{x^5}\,dx$

24) $\displaystyle\int \dfrac{-2x^2-4x-9}{x^4}\,dx$

25) $\int 4x(6x^4 + x^2 + 2)\, dx$

26) $\int \dfrac{-3x^5 - 3x + 8}{x^3}\, dx$

27) $\int \left(5x^4 - \dfrac{25\sqrt[3]{x^2}}{3} + 4\sqrt[3]{x}\right) dx$

28) $\int \left(5\sqrt[4]{x} - \dfrac{4}{x^2} - \dfrac{6}{x^4}\right) dx$

29) $\int \dfrac{7x^4 x^{\frac{3}{4}} + 8x^2 - 36}{4x^4}\, dx$

30) $\int \dfrac{12x^4 \sqrt[5]{x} - 40x - 45}{5x^4}\, dx$

Integrals Using u-Substitution

Notice that the reverse power rule is written in terms of *u* and *du* instead of *x* and *dx*. This is because not only can we integrate powers of x, but we can integrate powers of any function. We can integrate composition of functions using u-substitution, which is basically the chain rule for integration.

u-substitution works by noticing a composition of functions. Then we assign *u* to be the innermost function. We find the derivative of our chosen *u* which will be *du*. Usually *du* will also be somewhere in the expression, or we may need some algebraic manipulation to induce its presence.

<u>Ex. 1</u> Evaluate $\int 8x(4x^2 - 5)^5 \, dx$

<u>Solution:</u> Notice we have a composition of functions. We could expand out the expression $8x(4x^2 - 5)^5$, but this would be quite tedious. Instead we use the fact that it is a composition and use u-substitution. What happens if $u = 4x^2 - 5$? Then we would have $du = 8x \, dx$. This is great because our expression already has a this. Here's the substitution:

$$\int 8x(4x^2 - 5)^5 \, dx = \int \underbrace{(4x^2 - 5)^5}_{u} \underbrace{8x \, dx}_{du}$$

$$\Downarrow$$

$$\int u^5 \, du$$

Now we simply integrate $\int u^5 \, du = \frac{u^6}{6} + C$.

But wait, this expression is now in terms of u and our original problem was expressed in x's. So, we must go back to x: $\frac{u^6}{6} + C = \frac{1}{6}(4x^2 - 5)^6 + C$. Therefore, $\int 8x(4x^2 - 5)^5 \, dx = \frac{1}{6}(4x^2 - 5)^6 + C$.

Ex. 2 Evaluate $\int (5x^5 + 3)^3 \cdot 50x^4 \, dx$

Solution: Here we have a composition of functions. We will try u-substitution allowing $u = 5x^5 + 3$. This gives $du = 25x^4 \, dx$. We don't have exactly that in our problem so we need to make an adjustment by a constant: $\int (5x^5 + 3)^3 \cdot 50x^4 \, dx = 2 \int (5x^5 + 3)^3 \cdot 25x^4 \, dx$ Now it is set up to substitute:

$$2 \int (5x^5 + 3)^3 \cdot 25x^4 \, dx = 2 \int (u)^3 du = 2\frac{u^4}{4} + C = \frac{1}{2}(5x^5 + 3)^4 + C$$

Therefore, $\int (5x^5 + 3)^3 \cdot 50x^4 \, dx = \frac{1}{2}(5x^5 + 3)^4 + C$

Ex. 3 Evaluate $\int x(x - 4)^5 \, dx$

Solution: Here again we have a composition of functions, so u-substitution seems the way to go. $u = x - 4$ which implies $du = dx$. It seems we have a problem, because we don't know what to do with the extra x on the outside of $(x-4)^5$. This requires a little trick. We can rewrite that extra x in terms of u: $u = x - 4$ implies that $u + 4 = x$.

$$\int \underbrace{x}_{u+4} \underbrace{(x-4)^5}_{u} \underbrace{dx}_{du} \Rightarrow \int (u+4)u^5 \, du \Rightarrow \int (u^6+4u^5) \, du$$

$$\int (u^6 + 4u^5) \, du = \frac{u^7}{7} + \frac{4u^6}{6} + C = \frac{1}{7}(x-4)^7 + \frac{2}{3}(x-4)^6 + C$$

Therefore, $\int x(x-4)^5 \, dx = \frac{1}{7}(x-4)^7 + \frac{2}{3}(x-4)^6 + C$

Practice 40

Evaluate each indefinite integral. Use the provided substitution.

1) $\int 6x(3x^2 + 1)^5 \, dx; \; u = 3x^2 + 1$

2) $\int (2x^4 + 1)^4 \cdot 8x^3 \, dx; \; u = 2x^4 + 1$

Evaluate each indefinite integral.

3) $\int 25x^4(5x^5 - 3)^5 \, dx$

4) $\int (3x^5 + 5)^4 \cdot 15x^4 \, dx$

5) $\int 9x^2(3x^3 + 5)^{-4} \, dx$

6) $\int 6x^2(2x^3 + 3)^{-3} \, dx$

7) $\displaystyle\int \frac{20x^4}{(4x^5-3)^5}\,dx$

8) $\displaystyle\int \frac{4x}{(2x^2+5)^4}\,dx$

9) $\displaystyle\int 9x^2(3x^3-5)^{\frac{3}{2}}\,dx$

10) $\displaystyle\int (2x^2+1)^{\frac{2}{3}}\cdot 4x\,dx$

11) $\displaystyle\int 4x\sqrt[3]{2x^2+3}\,dx$

12) $\displaystyle\int 12x^3\sqrt[3]{3x^4-1}\,dx$

13) $\displaystyle\int (5x^4-2)^3\cdot 100x^3\,dx$

14) $\displaystyle\int (2x^2+5)^4\cdot 8x\,dx$

15) $\int 32x\sqrt[3]{4x^2+3}\ dx$

16) $\int 40x^3\sqrt[3]{5x^4+1}\ dx$

17) $\int x(3x-2)^3\ dx$

18) $\int 2x(3x-1)^3\ dx$

19) $\int \dfrac{5x}{(x-3)^5}\ dx$

20) $\int 5x\sqrt[3]{4x+5}\ dx$

21) $\int 6e^{3x}\cdot(e^{3x}-5)^4\ dx$

22) $\int (e^{3x}+4)^3\cdot 12e^{3x}\ dx$

23) $\displaystyle\int \frac{4(2+\ln-3x)^3}{x}\,dx$

24) $\displaystyle\int 10\cos 5x \cdot \sin^5 5x\,dx$

25) $\displaystyle\int 8\sin-4x \cdot \cos^3-4x\,dx$

26) $\displaystyle\int 4\cos 2x \cdot \sin^5 2x\,dx$

27) $\displaystyle\int \frac{15e^{5x}}{\left(e^{5x}+2\right)^5}\,dx$

28) $\displaystyle\int -12\csc 3x\cot 3x \cdot (\csc 3x)^{\frac{2}{3}}\,dx$

29) $\displaystyle\int -4\csc^2 2x\sqrt{\cot 2x}\,dx$

30) $\displaystyle\int \frac{5}{x(-2+\ln 3x)^5}\,dx$

Integrals Involving e^x and $\frac{1}{x}$

Recall the derivatives for exponential and logarithmic functions, now we will introduce the formulas for the integrals involving these functions.

$$\int e^u du = e^u + C$$

$$\int a^u du = \frac{a^u}{\ln a} + C$$

$$\int \frac{1}{u} du = \ln|x| + C$$

Notice for the last formula, that we have the absolute value of x, because the domain of natural log is only for nonnegative values.

<u>Ex. 1</u> Solve $\int -3x^{-1} dx = \int -\frac{3}{x} dx = -3 \int \frac{1}{x} dx = -3 \ln|x| + C$

<u>Solution:</u> The solution is quite straightforward using the formula above: $\int -3x^{-1} dx = \int -\frac{3}{x} dx = -3 \int \frac{1}{x} dx = -3 \ln|x| + C$

<u>Ex. 2</u> Solve $\int 10x^4 e^{x^5-2} dx$

<u>Solution:</u> This integral appears to be more complicated, so u-substitution seems to be the way to go. But what do we pick for u? Notice that we have a polynomial multiplied by an exponential. The exponential is also a composition of functions. The inner function of the composition is $x^5 - 2$, so this seems like a good choice for u. Differentiate to find du = $5x^4$.

Notice that this is just a constant multiple of what we have in front of the expression.

$\int 10x^4 e^{x^5-2} dx$ Let $u = x^5 - 2, du = 5x^4 dx$. This gives $2 \int e^{x^5-2} \cdot 5x^4 dx =$

$2 \int e^u \cdot du = 2e^u + C$ Now rewrite back in terms of x to get our final answer: $2e^{x^5-2} + C$.

Practice 41

1) $\int -e^x \, dx$

2) $\int -5e^x \, dx$

3) $\int 4e^x \, dx$

4) $\int 2e^x \, dx$

5) $\int 2 \cdot 3^x \, dx$

6) $\int -5 \cdot 4^x \, dx$

7) $\int -2 \cdot 2^x \, dx$

8) $\int -2 \cdot 3^x \, dx$

9) $\displaystyle\int -4x^{-1}\,dx$

10) $\displaystyle\int -\frac{1}{x}\,dx$

11) $\displaystyle\int -\frac{4}{x}\,dx$

12) $\displaystyle\int \frac{3}{x}\,dx$

13) $\displaystyle\int 20x^3 e^{x^4-3}\,dx$

14) $\displaystyle\int -18x^2 e^{2x^3-5}\,dx$

15) $\displaystyle\int -\frac{6x}{x^2-2}\,dx$

16) $\displaystyle\int -\frac{100x^4}{5x^5+2}\,dx$

17) $\displaystyle\int \frac{9x^2}{x^3-4}\,dx$

18) $\displaystyle\int -20x^4 e^{x^5-3}\,dx$

19) $\displaystyle\int -\frac{5\sin 5x}{\cos 5x}\,dx$

20) $\displaystyle\int -\frac{3\csc^2 x}{\cot x}\,dx$

21) $\displaystyle\int \frac{8\sec^2 4x}{\tan 4x}\,dx$

22) $\displaystyle\int -\frac{8\cos 2x}{\sin 2x}\,dx$

23) $\displaystyle\int -\frac{15e^{3x}}{e^{3x}-2}\,dx$

24) $\displaystyle\int -\frac{3e^x}{e^x+5}\,dx$

25) $\displaystyle\int -\frac{15e^{5x}}{e^{5x}+3}\,dx$

26) $\displaystyle\int -\frac{12e^{3x}}{e^{3x}-3}\,dx$

27) $\int -\dfrac{3}{x(-2 + \ln 2x)}\,dx$

28) $\int \dfrac{2}{x(-2 + \ln -2x)}\,dx$

29) $\int -\dfrac{5}{x(5 + \ln -x)}\,dx$

30) $\int \dfrac{3}{x(-3 + \ln -3x)}\,dx$

Integrals Involving Trigonometric Functions

Now we will introduce trigonometric integrals which should be recognized as the reverse of the derivative formulas.

$\int \cos u\, du = \sin u + C$ $\int \sin u\, du = -\cos u + C$

$\int \sec^2 u\, du = \tan u + C$ $\int \csc^2 u\, du = -\cot u + C$

$\int \sec u\, \tan u\, du = \sec u + C$ $\int \csc u\, \cot u\, du = -\csc u + C$

Ex. 1 Solve $\int -27x^2 \cos(3x^3 - 1)\, dx$

Solution: This integral is a polynomial multiplied by a cosine function. The cosine function is a composition with the inner function being $3x^3 - 1$. We will use u-substitution. Let $u = 3x^3 - 1$, $du = 9x^2\, dx$. $\int -27x^2 \cos(3x^3 - 1)\, dx = -3\int \cos(3x^3 - 1) \cdot 9x^2\, dx = -3\int \cos u\, du$

Now integrate: $-3\int \cos u\, du = -3\sin u + C$ Substituting x back in we get: $-3\sin u + C = -3\sin(3x^3 - 1) + C$. Therefore, $\int -27x^2 \cos(3x^3 - 1)\, dx = -3\sin(3x^3 - 1) + C$.

Ex. 2 Solve $\int 5\tan x\, dx$

Solution: This integral seems fairly simple, but we don't have an integral formula for tan. Sometimes we have to do some trig and algebraic manipulation to get an integral into a form that we can integrate. Here we will rewrite tan in terms of sine and cosine: $\int 5\tan x\, dx = 5\int \tan x\, dx = 5\int \frac{\sin x}{\cos x}\, dx$. Now we will use u-substitution, but we need to make a choice for u. If we let $u = \sin x$, $du = \cos x\, dx$, but this is a problem because we have cos in the denominator. Let's try another let $u = \cos x$, $du = -\sin x\, dx$. Then: $5\int \frac{\sin x}{\cos x}\, dx = -5\int \frac{1}{\cos x} \cdot -\sin x\, dx = -5\int \frac{1}{u}\, du = -5\ln|u| + C$ Substituting x back in gives: $-5\ln|u| + c = -5\ln|\cos x| + c$ Properties of logs allows us to write this as $-5\ln|\cos x| + C = 5\ln\left|\frac{1}{\cos x}\right| + C = 5\ln|\sec x| + C$. Therefore, $\int 5\tan x\, dx = 5\ln|\sec x| + C$.

Practice 42

1) $\int 4\sin x \, dx$

2) $\int 3\cos x \, dx$

3) $\int -5\sin x \, dx$

4) $\int -3\cos x \, dx$

5) $\int 3\sec^2 x \, dx$

6) $\int \csc^2 x \, dx$

7) $\int -2\csc^2 x \, dx$

8) $\int 4\sec^2 x \, dx$

9) $\displaystyle\int -4\csc x\cot x\,dx$

10) $\displaystyle\int -5\sec x\tan x\,dx$

11) $\displaystyle\int 4\sec x\tan x\,dx$

12) $\displaystyle\int 2\csc x\cot x\,dx$

13) $\displaystyle\int 8x\csc^2(x^2+1)\,dx$

14) $\displaystyle\int -24x^2\sin(4x^3-1)\,dx$

15) $\displaystyle\int 40x^4\csc(2x^5+5)\cot(2x^5+5)\,dx$

16) $\displaystyle\int -4x\sec^2(x^2+4)\,dx$

17) $\int 48x^3 \cos(3x^4 + 4)\, dx$

18) $\int 100x^3 \sin(5x^4 + 1)\, dx$

19) $\int 60x^2 \sec(5x^3 - 4)\tan(5x^3 - 4)\, dx$

20) $\int 10x \sec^2(x^2 + 5)\, dx$

21) $\int 8\csc 4x \cot 4x \csc^2(\csc 4x)\, dx$

22) $\int -12e^{4x} \cos(e^{4x} - 5)\, dx$

23) $\int -\dfrac{4\csc^2(-3 + \ln x)}{x}\, dx$

24) $\int 3\sec^2 3x \cos(\tan 3x)\, dx$

25) $\int -4\cot x \, dx$

26) $\int 3\csc x \, dx$

27) $\int 3\sec x \, dx$

28) $\int -\tan x \, dx$

29) $\int -\dfrac{4}{\csc x} \, dx$

30) $\int \dfrac{4}{\sin^2 x} \, dx$

31) $\int \dfrac{3}{\cos^2 x} \, dx$

32) $\int -\dfrac{4}{\sec x} \, dx$

33) $\int 3e^{3x}\sec(e^{3x}-4)\tan(e^{3x}-4)\,dx$

34) $\int -\dfrac{2\sin 2x}{\sec(\cos 2x)}\,dx$

35) $\int 6\csc 3x\cot 3x\cos(\csc 3x)\,dx$

36) $\int -\dfrac{2x}{\cos^2(x^2+5)}\,dx$

37) $\int -\dfrac{10\sin -2x\cos(\cos -2x)}{\sin(\cos -2x)}\,dx$

38) $\int -32x^3\csc(2x^4-5)\cot(2x^4-5)\,dx$

39) $\int \dfrac{2\sec(4+\ln 4x)\tan(4+\ln 4x)}{x}\,dx$

40) $\int \dfrac{15e^{5x}}{\sin(e^{5x}-1)}\,dx$

Definite Integrals

To find definite integrals we have to find the antiderivative as we have been doing. Then we evaluate at the upper bound and lower bound, taking the difference between the two. When evaluating a definite integral we will have a numerical answer that represents the area under the curve between the two bounds. We don't have to worry about the + C.

Ex. 1 Find $\int_2^5 \left(-\frac{x^2}{2} + 4x - 5\right) dx$

Solution: Integrate to get $\int_2^5 \left(-\frac{x^2}{2} + 4x - 5\right) dx = [-\frac{x^3}{6} + 2x^2 - 5x]_2^5 = \left[-\frac{(5)^3}{6} + 2(5)^2 - 5(5)\right] - \left[-\frac{(2)^3}{6} + 2(2)^2 - 5(2)\right] = \frac{25}{6} - \left(-\frac{10}{3}\right) = \frac{15}{2} = 7.5$

Ex. 2 Find $\int_{-\frac{\pi}{4}}^{\frac{\pi}{4}} 2 \sec^2 x \, dx$

Solution: Integrate by first taking the two outside: $\int_{-\frac{\pi}{4}}^{\frac{\pi}{4}} 2 \sec^2 x \, dx = 2 \int_{-\frac{\pi}{4}}^{\frac{\pi}{4}} \sec^2 x \, dx$. Then integrate using our trig formula: $2 \int_{-\frac{\pi}{4}}^{\frac{\pi}{4}} \sec^2 x \, dx = 2[\tan x]_{-\frac{\pi}{4}}^{\frac{\pi}{4}} = 2\left[\left(\tan\frac{\pi}{4}\right) - \left(\tan\left(-\frac{\pi}{4}\right)\right)\right] = 2[1 - (-1)] = 2(2) = 4$.

Practice 43

1) $\int_{-4}^{-1} (2x+2)\, dx$

2) $\int_{-4}^{1} (x-2)\, dx$

3) $\int_{-2}^{3} \left(-\dfrac{x^2}{2} + 2x + 2\right) dx$

4) $\int_{-1}^{3} (-x^2 + 2x - 3)\, dx$

5) $\int_{2}^{3} (x^3 - 4x^2 + 5x)\, dx$

6) $\int_{-5}^{-2} (x^3 + 11x^2 + 40x + 46)\, dx$

7) $\int_{-1}^{1} (x^5 - 3x^3)\, dx$

8) $\int_{-1}^{1} (x^5 - 4x^3 + 3x - 2)\, dx$

9) $\displaystyle\int_{1}^{6} (2x-6)^{\frac{1}{3}}\, dx$

10) $\displaystyle\int_{3}^{4} 5(2x-6)^{\frac{1}{3}}\, dx$

11) $\displaystyle\int_{2}^{4} \frac{1}{x^3}\, dx$

12) $\displaystyle\int_{-2}^{1} \frac{5}{(x-3)^2}\, dx$

13) $\displaystyle\int_{1}^{4} \frac{1}{x+1}\, dx$

14) $\displaystyle\int_{-4}^{-3} -\frac{3}{x+2}\, dx$

15) $\displaystyle\int_{-2}^{1} e^{x}\, dx$

16) $\displaystyle\int_{1}^{3} e^{x-3}\, dx$

17) $\displaystyle\int_{-\pi}^{0} \sin x\, dx$

18) $\displaystyle\int_{\frac{\pi}{2}}^{\frac{3\pi}{4}} -2\csc x\cot x\, dx$

19) $\displaystyle\int_{-\frac{\pi}{2}}^{0} -\cos x\, dx$

20) $\displaystyle\int_{\frac{\pi}{3}}^{\frac{3\pi}{4}} 2\csc^2 x\, dx$

Integrals Involving Inverse Trigonometric Functions

Now we will go over the formulas for integrating inverse trig functions

$$\int \frac{1}{\sqrt{a^2 - u^2}}\, du = \sin^{-1}\frac{u}{a} + C$$

$$\int \frac{1}{u^2 + a^2}\, du = \frac{1}{a}\tan^{-1}\frac{u}{a} + C$$

$$\int \frac{1}{u\sqrt{u^2 - a^2}}\, du = \frac{1}{a}\sec^{-1}\frac{|u|}{a} + C$$

Practice 44

1) $\displaystyle\int \frac{1}{\sqrt{9-x^2}}\,dx$

2) $\displaystyle\int \frac{1}{\sqrt{25-x^2}}\,dx$

3) $\displaystyle\int \frac{1}{9+x^2}\,dx$

4) $\displaystyle\int \frac{1}{1+x^2}\,dx$

5) $\displaystyle\int \frac{1}{x\sqrt{x^2-4}}\,dx$

6) $\displaystyle\int \frac{1}{x\sqrt{x^2-1}}\,dx$

7) $\displaystyle\int \frac{10x}{5x^2\sqrt{25x^4-9}}\,dx;\ u=5x^2$

8) $\displaystyle\int \frac{6x^2}{4+4x^6}\,dx;\ u=2x^3$

9) $\displaystyle\int \frac{8x}{\sqrt{25-16x^4}}\,dx$

10) $\displaystyle\int \frac{10x^4}{16+4x^{10}}\,dx$

11) $\displaystyle\int \frac{6x^2}{2x^3\sqrt{4x^6-4}}\,dx$

12) $\displaystyle\int \frac{4x^3}{16+x^8}\,dx$

13) $\displaystyle\int \frac{12x^2}{4x^3\sqrt{16x^6-16}}\,dx$

14) $\displaystyle\int \frac{6x^2}{\sqrt{9-4x^6}}\,dx$

15) $\displaystyle\int \frac{3e^{3x}}{\sqrt{1-e^{6x}}}\,dx$

16) $\displaystyle\int \frac{1}{x\cdot \ln -3x \cdot \sqrt{(\ln -3x)^2-4}}\,dx$

17) $\displaystyle\int \frac{1}{x\sqrt{9-(\ln x)^2}}\,dx$

18) $\displaystyle\int \frac{1}{x(1+(\ln -2x)^2)}\,dx$

19) $\displaystyle\int \frac{1}{x\cdot \ln -3x \cdot \sqrt{(\ln -3x)^2 - 16}}\,dx$

20) $\displaystyle\int \frac{e^x}{\sqrt{1-e^{2x}}}\,dx$

SOLUTIONS

Solutions Practice 1

1) 7
2) −7
3) $\sqrt{7}$
4) −3
5) −1
6) 4
7) $\dfrac{1}{2}$
8) 2
9) 12
10) 9
11) 18
12) 3
13) 4
14) 2

Solutions Practice 2

1) 13
2) Limit does not exist
3) 6
4) 8
5) ∞ or does not exist
6) C
7) 3
8) 2
9) The limit does not exist.
10) 1
11) 3
12) −∞ or The limit does not exist.

Solutions Practice 3

1) 6
2) 3
3) $-\dfrac{1}{2}$
4) $-\dfrac{1}{4}$
5) 6
6) $\dfrac{1}{8}$
7) $\dfrac{1}{2}$
8) 6
9) −4
10) −1
11) $-\dfrac{1}{36}$
12) $-\dfrac{1}{25}$
13) 0
14) 0
15) $\dfrac{1}{2}$
16) $\sqrt{2}$
17) 2
18) $\dfrac{1}{2}$
19) 4
20) 3

Solutions Practice 4

1) 5
2) 7
3) D
4) A
5) 0
6) 1
7) $\dfrac{5}{4}$
8) $\dfrac{1}{3}$
9) $-\dfrac{4}{25}$
10) $\dfrac{5}{2}$

Solutions Practice 5

1) jump discontinuity at $x = -1$
2) oscillating discontinuity at $x = 0$
3) removable discontinuity at $x = 1$
4) infinite discontinuity at $x = -2$
5) 3
6) Does not exist.
7) $(-\infty, -3), (-3, -1), (-1, \infty)$
8) 3
9) There is a vertical asymptote at $x = 2$, and horizontal asymptote at $y = 6$.
10) 6
11) 5
12) $\lim\limits_{x \to 3^-} g(x) = -\infty$
13) -2
    $\lim\limits_{x \to 3^+} g(x) = \infty$
14) 1
15) 2
16) 0
17) Does not exist.
18) Does not exist.
19) $-\dfrac{1}{2}$
20) Does not exist.
21) A
22) B
23) C
24) C

Solutions Practice 6

1. There exists some x in the interval $[-6, -4]$ such that $f(x) = -4$ NO

2. There exists some x in the interval $[-4, 0]$ such that $f(x) = 6$ YES

3. There exists some x in the interval $[4,6]$ such that $f(x) = 0$ YES

4. There exists at least one zero in the interval $[2,4]$ NO

5. There exists some x in the interval $[-2, 2]$ such that $f(x) = -6$ NO

6. What is the minimum number of zeros between $[-6, 6]$ and where do they occur? 4 at

    (-4, -2), (-2, 0), (0, 2), (4, 6)

Solutions Practice 7

1) 2
2) $\dfrac{1}{10}$
3) $y = -2x - 1$
4) $y = -\dfrac{1}{6}x + \dfrac{1}{2}$
5) $g'(1) > g'(4)$
6) $h'(-3) < h'(-2)$
7) B
8) D
9) 3
10) -3
11) 2
12) 1

Solutions Practice 8

1) $f'(x) = 2$
2) $f'(x) = 5$
3) $f'(x) = 3$
4) $f'(x) = -4$
5) $f'(x) = 6x$
6) $f'(x) = -4x$
7) $f'(x) = 10x$
8) $f'(x) = 8x$
9) $f'(x) = 10x + 1$
10) $f'(x) = 6x + 2$
11) $f'(x) = 8x + 4$
12) $f'(x) = 2x + 3$
13) $f'(x) = \dfrac{1}{2\sqrt{x+5}}$
14) $f'(x) = \dfrac{1}{\sqrt{2x+1}}$
15) $f'(x) = \dfrac{1}{2\sqrt{x-5}}$
16) $f'(x) = -\dfrac{3}{2\sqrt{-3x+1}}$
17) $f'(x) = -\dfrac{1}{x^2 - 8x + 16}$
18) $f'(x) = \dfrac{2}{x^2 - 2x + 1}$
19) $f'(x) = -\dfrac{4}{4x^2 - 12x + 9}$
20) $f'(x) = \dfrac{4}{4x^2 + 12x + 9}$

Solutions Practice 9

1) nondifferentiable; corner
2) differentiable
3) nondifferentiable; vertical tangent
4) nondifferentiable; discontinuous
5) nondifferentiable; cusp
6) differentiable
7) discontinuity, vertical asymptote
8) cusp
9) vertical tangent

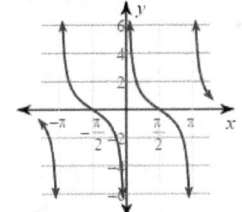
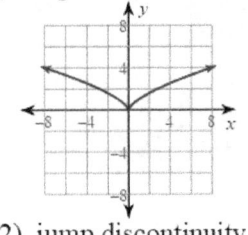
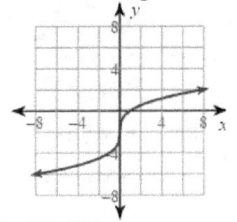

10) differentiable
11) corner
12) jump discontinuity
13) TRUE

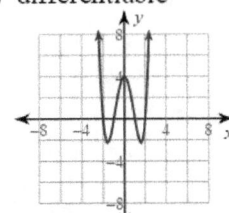
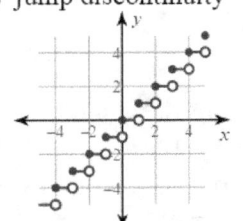

14) FALSE
15) FALSE
16) TRUE
17) FALSE
18) TRUE
19) TRUE
20) FALSE
21) TRUE
22) FALSE

Solutions Practice 10

1) $f'(x) = 25x^4 - 3x^2$ 2) $f'(x) = -12x^3 + 10x$ 3) $\dfrac{dy}{dx} = 8x^3 - \dfrac{8}{x^3} - \dfrac{8}{x^5}$

4) $\dfrac{dy}{dx} = -5x^4 - \dfrac{6}{x^4}$ 5) $f'(x) = -\dfrac{2}{x^2} - \dfrac{8}{x^5}$ 6) $f'(x) = -\dfrac{2}{x^3} - \dfrac{20}{x^6}$

7) $f'(x) = \dfrac{25x^{\frac{2}{3}}}{3} + \dfrac{6}{5x^{\frac{2}{5}}}$ 8) $\dfrac{dy}{dx} = \dfrac{4}{x^{\frac{1}{5}}} + \dfrac{4}{5x^{\frac{4}{5}}}$ 9) $f'(x) = \dfrac{1}{3x^{\frac{2}{3}}} + \dfrac{1}{5x^{\frac{4}{5}}}$

10) $\dfrac{dy}{dx} = \dfrac{2}{3x^{\frac{1}{3}}} + \dfrac{2}{3x^{\frac{2}{3}}} - \dfrac{2}{5x^{\frac{4}{5}}}$ 11) $\dfrac{dy}{dx} = -\dfrac{4}{3x^{\frac{1}{3}}} - \dfrac{1}{x^{\frac{2}{3}}} - \dfrac{16}{x^5}$ 12) $\dfrac{dy}{dx} = \dfrac{2}{x^{\frac{1}{2}}} - \dfrac{10}{x^3} - \dfrac{6}{x^4}$

13) $\dfrac{dy}{dx} = 5x^4 - 6x^{\frac{1}{2}} + \dfrac{10}{3x^{\frac{1}{3}}}$ 14) $\dfrac{dy}{dx} = -\dfrac{6}{5x^{\frac{2}{5}}} - \dfrac{2}{x^{\frac{1}{2}}} - \dfrac{10}{x^3}$ 15) $\dfrac{dy}{dx} = -20x^4 - 5x^{\frac{2}{3}} + \dfrac{2}{3x^{\frac{1}{3}}}$

16) $\dfrac{dy}{dx} = 2 + \dfrac{3}{4x^{\frac{1}{4}}} - \dfrac{12}{x^5}$

Solutions Practice 11

1) $f'(x) = -2x^3 \cdot 12x^2 + (4x^3 + 1) \cdot -6x^2$
   $= -48x^5 - 6x^2$

2) $\dfrac{dy}{dx} = 3x^2 \cdot 20x^4 + (4x^5 - 4) \cdot 6x$
   $= 84x^6 - 24x$

3) $\dfrac{dr}{ds} = (2s^4 + 4) \cdot -10s + (-5s^2 + 2) \cdot 8s^3$
   $= -60s^5 + 16s^3 - 40s$

4) $g'(x) = (5x^2 + 3) \cdot 4x^3 + (x^4 + 1) \cdot 10x$
   $= 30x^5 + 12x^3 + 10x$

5) $h'(w) = (-2w^2 - 3)(3w^2 + 6w) + (w^3 + 3w^2 - 2) \cdot -4w$
   $= -10w^4 - 24w^3 - 9w^2 - 10w$

6) $h'(x) = (-2x^4 + 2x^3 + 5) \cdot 16x^3 + (4x^4 + 1)(-8x^3 + 6x^2)$
   $= -64x^7 + 56x^6 + 72x^3 + 6x^2$

7) $\dfrac{dg}{dt} = (-4t^4 - 2)(-9t^2 - 8t) + (-3t^3 - 4t^2 + 1) \cdot -16t^3$
   $= 84t^6 + 96t^5 - 16t^3 + 18t^2 + 16t$

8) $\dfrac{df}{ds} = (3s^3 + 4s^2 + 4) \cdot -6s + (-3s^2 - 1)(9s^2 + 8s)$
   $= -45s^4 - 48s^3 - 9s^2 - 32s$

9) $f'(x) = 3x^{\frac{1}{4}} \cdot 16x^3 + (4x^4 - 5) \cdot \dfrac{3}{4}x^{-\frac{3}{4}}$
   $= 51x^{\frac{13}{4}} - \dfrac{15}{4x^{\frac{3}{4}}}$

10) $f'(x) = (1 + 3x^{-4}) \cdot -6x^2 - 2x^3 \cdot -12x^{-5}$
    $= -6x^2 + \dfrac{6}{x^2}$

Solutions Practice 12

1) $\dfrac{dy}{dx} = -\dfrac{4 \cdot 4x^3}{(x^4 + 5)^2}$
$= -\dfrac{16x^3}{x^8 + 10x^4 + 25}$

2) $f'(x) = -\dfrac{4 \cdot 8x}{(4x^2 - 3)^2}$
$= -\dfrac{32x}{16x^4 - 24x^2 + 9}$

3) $g'(r) = \dfrac{(4r^4 - 3) \cdot 10r^4 - (2r^5 + 4) \cdot 16r^3}{(4r^4 - 3)^2}$
$= \dfrac{8r^8 - 30r^4 - 64r^3}{16r^8 - 24r^4 + 9}$

4) $\dfrac{ds}{dr} = \dfrac{(5r^3 + 4)(25r^4 + 16r^3) - (5r^5 + 4r^4) \cdot 15r^2}{(5r^3 + 4)^2}$
$= \dfrac{50r^7 + 20r^6 + 100r^4 + 64r^3}{25r^6 + 40r^3 + 16}$

5) $\dfrac{ds}{dx} = \dfrac{(3x^2 + 4)(3x^2 + 8x) - (x^3 + 4x^2) \cdot 6x}{(3x^2 + 4)^2}$
$= \dfrac{3x^4 + 12x^2 + 32x}{9x^4 + 24x^2 + 16}$

6) $g'(w) = \dfrac{(5w^3 + 4) \cdot 3w^2 - (w^3 + 5) \cdot 15w^2}{(5w^3 + 4)^2}$
$= -\dfrac{63w^2}{25w^6 + 40w^3 + 16}$

7) $f'(x) = \dfrac{(2x^4 - 5)(10x^4 + 3x^2) - (2x^5 + x^3) \cdot 8x^3}{(2x^4 - 5)^2}$
$= \dfrac{4x^8 - 2x^6 - 50x^4 - 15x^2}{4x^8 - 20x^4 + 25}$

8) $\dfrac{dy}{dx} = \dfrac{(4x^2 + 2) \cdot 5x^4 - (x^5 + 5) \cdot 8x}{(4x^2 + 2)^2}$
$= \dfrac{6x^6 + 5x^4 - 20x}{8x^4 + 8x^2 + 2}$

9) $\dfrac{dy}{dx} = \dfrac{(3 + 3x^{-2}) \cdot 6x - (3x^2 + 5) \cdot -6x^{-3}}{(3 + 3x^{-2})^2}$
$= \dfrac{6x^5 + 12x^3 + 10x}{3x^4 + 6x^2 + 3}$

10) $\dfrac{dy}{dx} = \dfrac{\left(2x^{\frac{1}{3}} - 4\right) \cdot 8x - (4x^2 + 5) \cdot \dfrac{2}{3}x^{-\frac{2}{3}}}{\left(2x^{\frac{1}{3}} - 4\right)^2}$
$= \dfrac{20x^2 - 48x^{\frac{5}{3}} - 5}{6x^{\frac{4}{3}} - 24x + 24x^{\frac{2}{3}}}$

Solutions Practice 13

1) $5e^x$

2) $f'(x) = -3e^x$

3) $f'(x) = \dfrac{6}{x}$

4) $f'(x) = -\dfrac{5}{x}$

5) $f'(x) = 3^x \ln 3$

6) $f'(x) = 10^x \ln 10$

7) $f'(x) = \dfrac{6}{x \ln 4}$

8) $f'(x) = \dfrac{3}{x \ln 7}$

9) $f'(x) = \dfrac{1}{x \ln 10}$

10) $f'(x) = \dfrac{2}{x \ln 10}$

Solutions Practice 14

1) $2x + \cos x$

2) $\dfrac{1}{x^2} - 2\sin x$

3) $x(-2\cos x + x\sin x)$

4) $4 + \cot x - x\csc^2 x$

5) $x^2(x\sec^2 x + 3\tan x)$

6) $x^3(4\cot x - x\csc^2 x)$

7) $\dfrac{\sec x \cdot (1 + 2x\tan x)}{2\sqrt{x}}$

8) $\dfrac{(1 - 3x\cot x)\csc x}{3x^{\frac{2}{3}}}$

9) $-\dfrac{2\cos x + x\sin x}{x^3}$

10) $\dfrac{x\cos x - 2\sin x}{x^3}$

11) $-6\cot x \csc x$

12) $-4\csc^2 x$

13) $\dfrac{\cos x}{(1 - \sin x)^2}$

14) $-\dfrac{\sin x}{(1 - \cos x)^2}$

15) $2\cos x \sin x$

16) $-2\cos x \sin x$

17) -4

18) 1

19) $\dfrac{\sqrt{2}}{2}$

20) 0

21) $(\pi, 2)$
$y = 2$

22) $(0, 0)$
$y = -2x$

23) $\left(-\dfrac{\pi}{3}, -\dfrac{\sqrt{3}}{2}\right)$
$y = \dfrac{1}{2}x + \dfrac{-3\sqrt{3} + \pi}{6}$

24) $\left(\dfrac{\pi}{6}, -\dfrac{2\sqrt{3}}{3}\right)$
$y = -\dfrac{2}{3}x + \dfrac{-6\sqrt{3} + \pi}{9}$

Solutions Practice 15

1) $\dfrac{dy}{dx} = 4(4x^3 + 1)^3 \cdot 12x^2$
$= 48x^2(4x^3 + 1)^3$

2) $\dfrac{dy}{dx} = 3(2x^3 + 5)^2 \cdot 6x^2$
$= 18x^2(2x^3 + 5)^2$

3) $\dfrac{dy}{dx} = \dfrac{1}{3}(5x^3 - 4)^{-\frac{2}{3}} \cdot 15x^2$
$= \dfrac{5x^2}{(5x^3 - 4)^{\frac{2}{3}}}$

4) $\dfrac{dy}{dx} = \dfrac{1}{4}(x^5 + 5)^{-\frac{3}{4}} \cdot 5x^4$
$= \dfrac{5x^4}{4(x^5 + 5)^{\frac{3}{4}}}$

5) $\dfrac{dy}{dx} = -3(3x^2 + 1)^{-4} \cdot 6x$
$= -\dfrac{18x}{(3x^2 + 1)^4}$

6) $\dfrac{dy}{dx} = -2(5x^3 - 3)^{-3} \cdot 15x^2$
$= -\dfrac{30x^2}{(5x^3 - 3)^3}$

7) $\dfrac{dy}{dx} = 3((5x^4 + 3)^5 - 3)^2 \cdot 5(5x^4 + 3)^4 \cdot 20x^3$
$= 300x^3((5x^4 + 3)^5 - 3)^2 \cdot (5x^4 + 3)^4$

8) $\dfrac{dy}{dx} = 5((3x^5 - 2)^2 - 5)^4 \cdot 2(3x^5 - 2) \cdot 15x^4$
$= 150x^4((3x^5 - 2)^2 - 5)^4(3x^5 - 2)$

9) $\dfrac{dy}{dx} = \cos 3x^2 \cdot 6x$
$= 6x\cos 3x^2$

10) $\dfrac{dy}{dx} = -\sin 3x^5 \cdot 15x^4$
$= -15x^4 \sin 3x^5$

11) $\dfrac{dy}{dx} = e^{2x^4} \cdot 8x^3$

12) $\dfrac{dy}{dx} = e^{3x^4} \cdot 12x^3$ 13) $\dfrac{dy}{dx} = e^{x^2} \cdot 2x$ 14) $\dfrac{dy}{dx} = e^{3x^5} \cdot 15x^4$

15) $\dfrac{dy}{dx} = (x^5 + 5) \cdot 5(3x^4 - 2)^4 \cdot 12x^3 + (3x^4 - 2)^5 \cdot 5x^4$

$\phantom{15)\ \dfrac{dy}{dx}} = 5x^3(3x^4 - 2)^4(15x^5 + 60 - 2x)$

16) $\dfrac{dy}{dx} = (-2x^2 + 5) \cdot 3(2x + 1)^2 \cdot 2 + (2x + 1)^3 \cdot -4x$

$\phantom{16)\ \dfrac{dy}{dx}} = 2(2x + 1)^2(-10x^2 + 15 - 2x)$

17) $\dfrac{dy}{dx} = \dfrac{(4x^4 + 5)^2 \cdot 2 - (2x + 5) \cdot 2(4x^4 + 5) \cdot 16x^3}{((4x^4 + 5)^2)^2}$

$\phantom{17)\ \dfrac{dy}{dx}} = \dfrac{2(-28x^4 + 5 - 80x^3)}{(4x^4 + 5)^3}$

18) $\dfrac{dy}{dx} = \dfrac{(5x^3 + 4) \cdot 2(x^2 - 2) \cdot 2x - (x^2 - 2)^2 \cdot 15x^2}{(5x^3 + 4)^2}$

$\phantom{18)\ \dfrac{dy}{dx}} = \dfrac{x(x^2 - 2)(5x^3 + 16 + 30x)}{(5x^3 + 4)^2}$

19) $h_1'(3) = -\dfrac{1}{2}$

$h_2'(3) = -\dfrac{1}{2}$

$h_3'(2) = \dfrac{15}{2}$

$h_4'(2) = -\dfrac{1}{6}$

$h_5'(1) = 4$

$h_6'(3) = 0$

20) $h_1'(2) = \dfrac{1}{2}$

$h_2'(1) = 4$

$h_3'(4) = -2$

$h_4'(1) = \dfrac{10}{9}$

$h_5'(1) = 8$

$h_6'(1) = 2$

Solutions Practice 16

1) $\dfrac{dy}{dx} = -\dfrac{5x}{y}$ 2) $\dfrac{dy}{dx} = -\dfrac{2x}{5y^2}$ 3) $\dfrac{dy}{dx} = -\dfrac{9x^2}{4 - 2y}$ 4) $\dfrac{dy}{dx} = \dfrac{2x}{-1 + y}$

5) $\dfrac{dy}{dx} = \dfrac{-4y^3 - 3}{4xy^2}$ 6) $\dfrac{dy}{dx} = \dfrac{3x^2 - y^2}{2xy}$ 7) $\dfrac{dy}{dx} = \dfrac{5y^2 - 9x^2}{6y - 10yx}$ 8) $\dfrac{dy}{dx} = \dfrac{3 + 2xy^2}{4y - 2yx^2}$

9) $\dfrac{dy}{dx} = \dfrac{x^2}{10y^5 + 2y^2}$ 10) $\dfrac{dy}{dx} = \dfrac{2x^2}{y^5 + 2y^2}$ 11) $\dfrac{dy}{dx} = -\dfrac{5x}{4y\csc 4y^2 \csc 4y^2}$

12) $\dfrac{dy}{dx} = \dfrac{3x^2}{2y\cos 5y^2}$ 13) $\dfrac{dy}{dx} = \dfrac{x}{6y^2 e^{4y^3}}$ 14) $\dfrac{dy}{dx} = 3yx$ 15) $y = -\dfrac{1}{4}x + \dfrac{3}{2}$

16) $y = -\dfrac{3}{2}x + \dfrac{1}{2}$ 17) $y = -9x - 16$ 18) $y = \dfrac{1}{7}x + \dfrac{5}{7}$ 19) $\dfrac{d^2y}{dx^2} = -\dfrac{1}{9y^3}$

20) $\dfrac{d^2y}{dx^2} = \dfrac{60xy^2 - 36x^4}{25y^3}$ 21) $\dfrac{d^2y}{dx^2} = \dfrac{10y^2 - 25x^2}{4y^3}$ 22) $\dfrac{d^2y}{dx^2} = \dfrac{15y^2 - 9x^2}{25y^3}$

23) $\left.\dfrac{d^2y}{dx^2}\right|_{x=2, y=1} = -\dfrac{1}{4}$ 24) $\left.\dfrac{d^2y}{dx^2}\right|_{x=-1, y=3} = \dfrac{8}{3}$

Solutions Practice 17

1) $-\dfrac{1}{5}$ 2) $\dfrac{2}{3}$ 3) -3 4) $\dfrac{3}{5}$

5) $\dfrac{1}{2}$ 6) $\dfrac{1}{4}$ 7) $-\dfrac{1}{5}$ 8) 1

9) $-\dfrac{1}{9}$ 10) $\dfrac{1}{36}$ 11) 9 12) $\dfrac{27}{5}$

13) $\dfrac{1}{2}$ 14) $-\dfrac{6}{5}$ 15) $(f^{-1})'(x) = -\dfrac{x}{2}$ 16) $(f^{-1})'(x) = \dfrac{2x}{5}$

17) $(f^{-1})'(x) = \dfrac{3x^2}{5}$ 18) $(f^{-1})'(x) = -\dfrac{3x^2}{4}$ 19) $(f^{-1})'(x) = \dfrac{1}{6 \cdot \left(\dfrac{x-5}{3}\right)^{\frac{1}{2}}}$

20) $(f^{-1})'(x) = \dfrac{1}{6 \cdot \left(\dfrac{x-2}{2}\right)^{\frac{2}{3}}}$

Solutions Practice 18

1) $f'(x) = \dfrac{1}{\sqrt{1-(x^2)^2}} \cdot 2x = \dfrac{2x}{\sqrt{1-x^4}}$

2) $f'(x) = \dfrac{1}{\sqrt{1-(3x^5)^2}} \cdot 15x^4 = \dfrac{15x^4}{\sqrt{1-9x^{10}}}$

3) $f'(x) = -\dfrac{1}{\sqrt{1-(4x^3)^2}} \cdot 12x^2 = -\dfrac{12x^2}{\sqrt{1-16x^6}}$

4) $f'(x) = -\dfrac{1}{\sqrt{1-(5x^5)^2}} \cdot 25x^4 = -\dfrac{25x^4}{\sqrt{1-25x^{10}}}$

5) $f'(x) = \dfrac{1}{(5x^5)^2+1} \cdot 25x^4 = \dfrac{25x^4}{25x^{10}+1}$

6) $f'(x) = \dfrac{1}{(5x^3)^2+1} \cdot 15x^2 = \dfrac{15x^2}{25x^6+1}$

7) $f'(x) = -\dfrac{1}{(x^3)^2+1} \cdot 3x^2 = -\dfrac{3x^2}{x^6+1}$

8) $f'(x) = -\dfrac{1}{(3x^3)^2+1} \cdot 9x^2 = -\dfrac{9x^2}{9x^6+1}$

9) $f'(x) = \dfrac{1}{|3x^2|\sqrt{(3x^2)^2-1}} \cdot 6x = \dfrac{2}{x\sqrt{9x^4-1}}$

10) $f'(x) = \dfrac{1}{|2x^3|\sqrt{(2x^3)^2-1}} \cdot 6x^2 = \dfrac{6x^2}{|2x^3|\sqrt{4x^6-1}}$

11) $f'(x) = -\dfrac{1}{|-2x^5|\sqrt{(-2x^5)^2-1}} \cdot -10x^4 = \dfrac{10x^4}{|-2x^5|\sqrt{4x^{10}-1}}$

12) $f'(x) = -\dfrac{1}{|x^2|\sqrt{(x^2)^2-1}} \cdot 2x = -\dfrac{2}{x\sqrt{x^4-1}}$

13) $f'(x) = 5(\cot^{-1} 4x^3)^4 \cdot -\dfrac{1}{(4x^3)^2 + 1} \cdot 12x^2$

$= -\dfrac{60x^2(\cot^{-1} 4x^3)^4}{16x^6 + 1}$

14) $f'(x) = -\dfrac{1}{((-4x^3 + 5)^5)^2 + 1} \cdot 5(-4x^3 + 5)^4 \cdot -12x^2$

$= -\dfrac{60x^2(-4x^3 + 5)^4}{(-4x^3 + 5)^{10} + 1}$

15) $f'(x) = 4(\sec^{-1} -2x^2)^3 \cdot \dfrac{1}{|-2x^2|\sqrt{(-2x^2)^2 - 1}} \cdot -4x$

$= -\dfrac{16x(\sec^{-1} -2x^2)^3}{|-2x^2|\sqrt{4x^4 - 1}}$

16) $f'(x) = -\dfrac{1}{((4x^3 + 1)^5)^2 + 1} \cdot 5(4x^3 + 1)^4 \cdot 12x^2$

$= -\dfrac{60x^2(4x^3 + 1)^4}{(4x^3 + 1)^{10} + 1}$

17) $\dfrac{\sqrt{3}}{3}$ 18) $\dfrac{\sqrt{7}}{7}$ 19) $-\dfrac{\sqrt{7}}{7}$ 20) $-\dfrac{1}{4}$

21) $\dfrac{5}{29}$ 22) $\dfrac{3}{50}$ 23) $-\dfrac{2\sqrt{5}}{45}$ 24) $\dfrac{\sqrt{5}}{15}$

25) $-\dfrac{4}{25}$ 26) $-\dfrac{2\sqrt{5}}{15}$ 27) $-\dfrac{4}{25}$ 28) $-\dfrac{4}{25}$

Solutions Practice 19

1) $\dfrac{d^4y}{dx^4} = -240x - 72$ 2) $\dfrac{d^4y}{dx^4} = -480x - 24$ 3) $-\dfrac{15}{16x^{\frac{7}{2}}}$ 4) $-\dfrac{80}{81x^{\frac{11}{3}}}$

5) $32\cos 2x$ 6) $-243\sin 3x$ 7) $\tan x \sec x \cdot (\tan^2 x + 5\sec^2 x)$

8) $-\cot x \csc x \cdot (\cot^2 x + 5\csc^2 x)$ 9) $-2(\csc^4 x + 2\cot^2 x \csc^2 x)$

10) $2\sec^2 x(2\tan^2 x + \sec^2 x)$ 11) $81e^{3x}$ 12) $32e^{2x}$

13) $-\dfrac{6}{x^4}$ 14) $\dfrac{24}{x^5}$ 15) $\dfrac{27x}{(1 - 9x^2)^{\frac{3}{2}}}$ 16) $-\dfrac{64x}{(1 - 16x^2)^{\frac{3}{2}}}$

17) $\dfrac{16x}{(4x^2 + 1)^2}$ 18) $-\dfrac{54x}{(9x^2 + 1)^2}$ 19) $\dfrac{d^2y}{dx^2} = \dfrac{-6xy^2 - 9x^4}{y^3}$

20) $\dfrac{d^2y}{dx^2} = -\dfrac{25}{64y^3}$ 21) $\dfrac{d^2y}{dx^2} = \dfrac{6xy^2 - 9x^4}{y^3}$ 22) $\dfrac{d^2y}{dx^2} = \dfrac{12xy^2 - 9x^4}{4y^3}$

23) $f'(x) = \cos x$
$f''(x) = -\sin x$
$f'''(x) = -\cos x$
$f^{(4)}(x) = \sin x$
$f^{(67)}(x) = -\cos x$
$f^{(105)}(x) = \cos x$

24) $f'(x) = -\sin x$
$f''(x) = -\cos x$
$f'''(x) = \sin x$
$f^{(4)}(x) = \cos x$
$f^{(55)}(x) = \sin x$
$f^{(110)}(x) = -\cos x$

Solutions Practice 20

1) After 6 seconds, Ethan's acceleration is 2.5 meters per second squared.
2) When the weight is released, its height is increasing at a rate of 4 cm/s.
3) After 6 minutes, the pool is being filled at a rate of 0.2 meters per minute.
4) After 4 hours, the pond is losing water at a rate of 3 centimeters per hour.
5) When the ball is released, its velocity is decreasing at a rate of 10 m/s^2.
6) After 4 hours, the amount of medication is decreasing by 1.5 milligrams per hour.
7) On January 12, the day length is increasing at a rate of 4 minutes per day.
8) 2 hours after leaving, Matthew drove at a rate of 70 miles per hour.
9) 5 hours after leaving, Katie drove at a rate of 90 kilometers per hour.
10) After 10 years, the amount of money in increasing in the investment by $2000 per year.
11) 134 views/day   12) $9/kg   13) 1.36 degrees Celsius/hour
14) −120 Liters/minute

Solutions Practice 21

1) What is the position at $t = 7$? 4 m
What is the velocity at $t = 7$? −2 m/s
What is the speed at $t = 7$? 2 m/s
What is the position at $t = 14$? −1 m
What is the velocity at $t = 14$? $\frac{1}{2}$ m/s
What is the speed at $t = 14$? $\frac{1}{2}$ m/s
When is the particle moving right? (0, 6), (12, 18)
When is the particle moving left? (6, 10), (18, 20)
When is the particle at rest?
$t = 0, t = 6, (10, 12), t = 18$
When does the particle change direction? $t = 6, 12, 18$
What is the displacement? 1 m
What is the total distance traveled? 19 m

2) What is the position at $t = 4$? −2 m
What is the velocity at $t = 4$? 1 m/s
What is the position at $t = 14$? 6 m
What is the velocity at $t = 14$? $\frac{1}{2}$ m/s
What is the speed at $t = 4$? 1 m/s
What is the speed at $t = 17$? 1 m/s
When is the particle moving right? (2, 10), (12, 16)
When is the particle moving left? (0, 2), (16, 20)
When is the particle at rest?
$t = 2, (10, 12), t = 16$
When does the particle change direction?
$t = 2, t = 16$
What is the displacement?
3 m
What is the total distance traveled? 19 m

3) What is the velocity at $t = 3$? 0 m/s
   What is the acceleration at $t = 3$? 2 m/s^2
   What is the velocity at $t = 5$? 2 m/s
   What is the acceleration at $t = 5$? 0 m/s^2
   What is the velocity at $t = 12$? −4 m/s
   What is the acceleration at $t = 12$? 1 m/s^2
   What is the speed at $t = 8$? 2 m/s
   What is the speed at $t = 12$? 4 m/s
   When is the particle moving right?
   (3, 8.5), (15, 18)
   When is the particle moving left?
   (0, 3), (8.5, 15), (18, 20)
   When is the particle at rest?
   $t = 0, 3, 8.5, 15, 18$
   When does the particle change direction?
   $t = 3, 8.5, 15, 18$
   When is the particle speeding up?
   (0, 2), (3, 4), (8.5, 10), (15, 16), (18, 20)
   When is the particle slowing down?
   (2, 3), (8, 8.5), (10, 15), (16, 18)
5) moving to the right with constant velocity
7) moving to the left with constant velocity
9) moving to the right, speeding up
11) moving to the left, speeding up

4) What is the velocity at $t = 4$? 0 m/s
   What is the acceleration at $t = 4$? −2 m/s^2
   What is the velocity at $t = 7$? −4 m/s
   What is the acceleration at $t = 7$? 0 m/s^2
   What is the velocity at $t = 15$? 1 m/s
   What is the acceleration at $t = 15$? 1 m/s^2
   What is the speed at $t = 2$? 4 m/s
   What is the speed at $t = 12$? 2 m/s
   When is the particle moving right?
   (0, 4), (14, 17)
   When is the particle moving left?
   (4, 14), (17, 20)
   When is the particle at rest?
   $t = 4, 14, 17$
   When does the particle change direction?
   $t = 4, 14, 17$
   When is the particle speeding up?
   (4, 6), (14, 16), (17, 20)
   When is the particle slowing down?
   (2, 4), (8, 14), (16, 17)
6) moving to the left with constant velocity
8) Not moving
10) moving to the right, slowing down
12) moving to the left, slowing down

13) Answers may vary
14) Answers may vary
15) Answers may vary

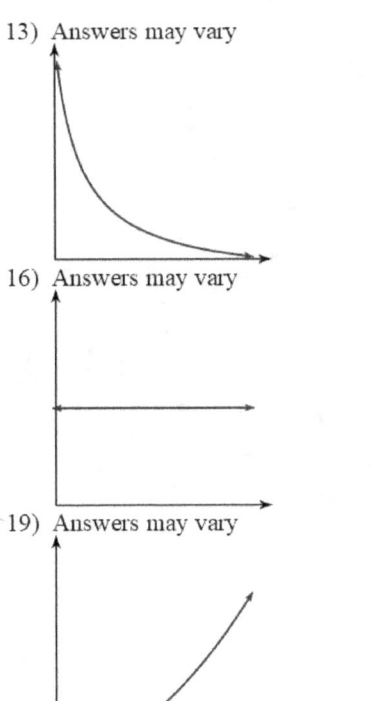
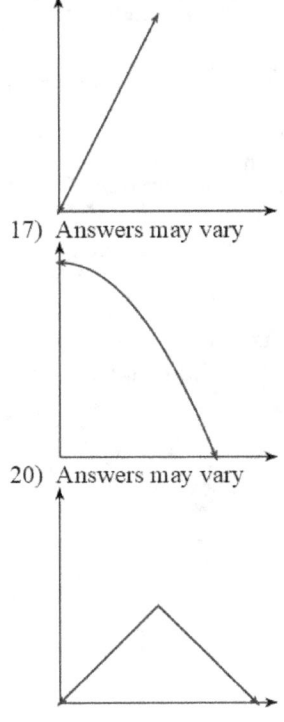
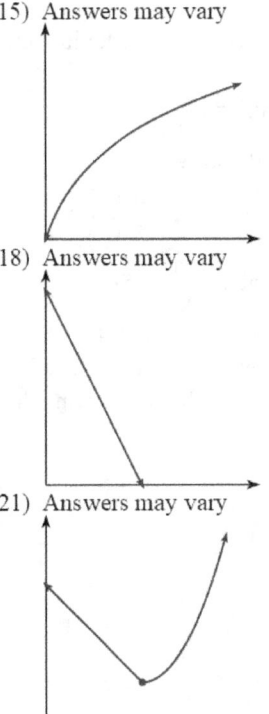

16) Answers may vary
17) Answers may vary
18) Answers may vary

19) Answers may vary
20) Answers may vary
21) Answers may vary

22) Answers may vary 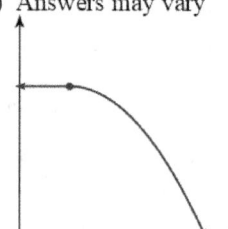 23) Answers may vary. 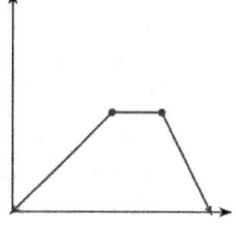 24) Answers may vary.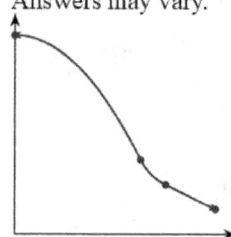

Solutions Practice 22

1) $s(5) = 30$, $v(5) = -1$, speed at 5 = 1, $a(5) = -2$
2) $s(8) = -3072$, $v(8) = -640$, speed at 8 = 640, $a(8) = 96$
3) $s(5) = 180$, $v(5) = -24$, speed at 5 = 24, $a(5) = -14$
4) $s(5) = -175$, $v(5) = -45$, speed at 5 = 45, $a(5) = 6$
5) $s(6) = -150$, $v(6) = 35$, speed at 6 = 35, $a(6) = 8$ 6) $s(3) = 180$, $v(3) = 9$, speed at 3 = 9, $a(3) = -28$
7) $v(t) = 2t - 16$, $a(t) = 2$
   Changes direction at: $t = \{8\}$, Moving left: $0 \le t < 8$, Moving right: $t > 8$
   Acceleration zero: Never, Slowing down: $0 \le t < 8$, Speeding up: $t > 8$
8) $v(t) = 2t - 25$, $a(t) = 2$
   Changes direction at: $t = \left\{\dfrac{25}{2}\right\}$, Moving left: $0 \le t < \dfrac{25}{2}$, Moving right: $t > \dfrac{25}{2}$
   Acceleration zero: Never, Slowing down: $0 \le t < \dfrac{25}{2}$, Speeding up: $t > \dfrac{25}{2}$
9) $v(t) = -3t^2 + 32t - 64$, $a(t) = -6t + 32$
   Changes direction at: $t = \left\{\dfrac{8}{3}, 8\right\}$, Moving left: $0 \le t < \dfrac{8}{3}$, $t > 8$, Moving right: $\dfrac{8}{3} < t < 8$
   Acceleration zero at: $t = \left\{\dfrac{16}{3}\right\}$, Slowing down: $0 \le t < \dfrac{8}{3}$, $\dfrac{16}{3} < t < 8$, Speeding up: $\dfrac{8}{3} < t < \dfrac{16}{3}$, $t > 8$
10) $v(t) = 3t^2 - 26t + 40$, $a(t) = 6t - 26$
    Changes direction at: $t = \left\{2, \dfrac{20}{3}\right\}$, Moving left: $2 < t < \dfrac{20}{3}$, Moving right: $0 \le t < 2$, $t > \dfrac{20}{3}$
    Acceleration zero at: $t = \left\{\dfrac{13}{3}\right\}$, Slowing down: $0 \le t < 2$, $\dfrac{13}{3} < t < \dfrac{20}{3}$, Speeding up: $2 < t < \dfrac{13}{3}$, $t > \dfrac{20}{3}$
11) $v(t) = -4t^3 + 24t^2$, $a(t) = -12t^2 + 48t$
    Changes direction at: $t = \{6\}$, Moving left: $t > 6$, Moving right: $0 < t < 6$
    Acceleration zero at: $t = \{0, 4\}$, Slowing down: $4 < t < 6$, Speeding up: $0 < t < 4$, $t > 6$

12) $v(t) = 4t^3 - 30t^2$, $a(t) = 12t^2 - 60t$
   Changes direction at: $t = \left\{\dfrac{15}{2}\right\}$, Moving left: $0 < t < \dfrac{15}{2}$, Moving right: $t > \dfrac{15}{2}$
   Acceleration zero at: $t = \{0, 5\}$, Slowing down: $5 < t < \dfrac{15}{2}$, Speeding up: $0 < t < 5$, $t > \dfrac{15}{2}$

13) Maximum speed: 8 at $t = \{10\}$
    Displacement: 12
    Distance traveled: 20

14) Maximum speed: 10 at $t = \{15\}$
    Displacement: −24
    Distance traveled: 26

15) Maximum speed: 40 at $t = \{0\}$
    Displacement: −30
    Distance traveled: 42

16) Maximum speed: 160 at $t = \{10\}$
    Displacement: 0
    Distance traveled: 576

17) Maximum speed: 250 at $t = \{5\}$
    Displacement: −960
    Distance traveled: $\dfrac{8171}{8} = 1021.375$

18) Maximum speed: 320 at $t = \{8\}$
    Displacement: 12
    Distance traveled: $\dfrac{47611}{128} \approx 371.961$

19) Maximum speed: 56 at $t = \{7\}$
    Displacement: −144
    Distance traveled: 144

20) Maximum speed: 60 at $t = \{6\}$
    Displacement: 75
    Distance traveled: $\dfrac{4375}{27} \approx 162.037$

Solutions Practice 23

1) $\dfrac{15}{4}$ ft/sec
2) $\dfrac{84}{5}$ ft/sec down the wall
3) −3 ft/sec
4) $\dfrac{25}{12}$ ft/sec
5) 360 ft/sec
6) $\dfrac{28}{625}$ radians/sec
7) 108π m²/min
8) 70π ft²/sec
9) −972 mm³/sec
10) 768 m³/min
11) 432π cm³/sec
12) -588π in³/sec
13) $-\dfrac{16}{147}$ cm/sec
14) 16π cm³/sec
15) 2 cm²/min
16) 176 cm²/s
17) 32 km²/s
18) 405 m³/s
19) $-\dfrac{30}{8}$ m/s
20) 10 km²/hr
21) 9π m³/s
22) 8 m/min
23) 32 m²/s
24) 16π m²/hr

Solutions Practice 24

1) −1
2) 25
3) 4
4) 12
5) 7.6
6) −1.4
7) 8
8) 6.1
9) B
10) D
11) B
12) C
13) A
14) C
15) D
16) C

Solutions Practice 25

1) $\dfrac{2}{5}$ 2) 1   3) ∞ 4) 0
5) 2   6) 4   7) ∞ 8) 2
9) 1   10) 2   11) 4   12) 0
13) -1 14) 3   15) 2   16) $\dfrac{5}{4}$
17) 1   18) 4   19) ∞ 20) 1
21) 1   22) $\dfrac{3}{4}$ 23) NO ∞ 24) YES 0
25) YES 3   26) NO ∞ 27) YES 3   28) NO ∞
29) 5   30) 0   31) 0   32) 2
33) 1   34) e^2 35) e 36) 5

Solutions Practice 26

1. Does there exist some x in the interval $[-6, -4]$ such that $f(x) = -2$?

   No, because -2 is not between -4 and -8

2. Does there exist some x in the interval $[4,6]$ such that $f(x) = -6$?

   Yes, because -6 is between -8 and -2

3. Does there exist some x in the interval $[-4,0]$ such that $f(x) = 2$?

   Yes, because 2 is between -8 and 4

4. Does there exist some x in the interval $[0,2]$ such that $f(x) = 2$?

   No, because 2 is not between 4 and 6

5. Does there exist some x in the interval $[6,10]$ such that $f(x) = 8$

   No, because 8 is not between -2 and 4

6. What is the minimum number of zeros $f(x)$ has? Where are there possible zeros?

   The minimum number of critical points, or zeros, are the number of sign changes which is 3 sign changes at (-4, 0), (2, 4) and (6,10).

Solutions Practice 27

1) $\{3\}$ 2) $\{-1\}$ 3) $\{-4\}$ 4) $\{-2\}$
5) $\left\{-\dfrac{1}{2}\right\}$ 6) $\left\{-\dfrac{5}{2}\right\}$ 7) $\left\{\dfrac{2}{3}\right\}$ 8) $\left\{\dfrac{5}{3}\right\}$
9) $\left\{\dfrac{3-2\sqrt{3}}{3}, \dfrac{3+2\sqrt{3}}{3}\right\}$ 10) $\left\{\dfrac{4-\sqrt{13}}{3}, \dfrac{4+\sqrt{13}}{3}\right\}$ 11) $\{-3\}$
12) $\{2\}$ 13) $\{-2+\sqrt{5}\}$ 14) $\{1-\sqrt{5}\}$ 15) $\{3\}$
16) $\left\{-\dfrac{15}{4}\right\}$ 17) The function is not continuous on $[1, 3]$ 18) $\left\{\dfrac{28}{9}\right\}$
19) $\{-\sqrt{3}\}$ 20) The function is not differentiable on $(3, 6)$

Solutions Practice 28

1) Absolute minimum: $(6, -5)$
   Absolute maximum: $(2, 3)$
2) Absolute minimum: $\left(-7, -\dfrac{7}{2}\right)$
   Absolute maximum: $\left(-5, \dfrac{1}{2}\right)$

3) Absolute minima: $(-2, -4), (0, -4)$
   Absolute maximum: $(-1, -2)$
4) Absolute minimum: $\left(-7, \dfrac{1}{2}\right)$
   Absolute maximum: $(-4, 5)$

5) Absolute minima: $(0, 3), (2, 3)$
   Absolute maximum: $\left(\dfrac{4}{3}, \dfrac{113}{27}\right)$
6) Absolute minima: $(0, 0), (3, 0)$
   Absolute maximum: $(2, 4)$

7) Absolute minimum: $(0, -2)$
   Absolute maximum: $(-1, 3)$
8) Absolute minima: $(1, 1), (0, 1)$
   Absolute maximum: $(-1, 3)$

9) Absolute minimum: $(0, -2)$
   Absolute maxima: $(-1, 1), (1, 1)$
10) Absolute minima: $(-1, -4), (1, -4), (0, -4)$
    Absolute maxima: $\left(-\dfrac{\sqrt{2}}{2}, -\dfrac{15}{4}\right), \left(\dfrac{\sqrt{2}}{2}, -\dfrac{15}{4}\right)$

11) Absolute minimum: $\left(\dfrac{\sqrt{6}}{2}, -\dfrac{9}{4}\right)$
    Absolute maximum: $(2, 4)$
12) Absolute minima: $(-2, -5), (0, -5)$
    Absolute maximum: $(-\sqrt{2}, -1)$

13) Absolute minimum: $\left(-4, \dfrac{1}{17}\right)$
    Absolute maximum: $\left(-2, \dfrac{1}{5}\right)$
14) Absolute minima: $\left(-2, -\dfrac{1}{6}\right), \left(2, -\dfrac{1}{6}\right)$
    Absolute maximum: $\left(0, -\dfrac{1}{8}\right)$

15) Absolute minimum: $\left(2, \dfrac{8}{5}\right)$
    Absolute maximum: $(4, 2)$

16) Absolute minimum: $(0, -5)$
    Absolute maxima: $\left(-2, -\dfrac{5}{2}\right), \left(2, -\dfrac{5}{2}\right)$

17) Absolute minimum: $\left(-5, \sqrt[3]{6}\right)$
    Absolute maximum: $\left(1, \sqrt[3]{42}\right)$

18) Absolute minimum: $\left(-3, 2\sqrt{3}\right)$
    Absolute maximum: $\left(3, 4\sqrt{3}\right)$

19) Absolute minimum: $\left(0, -3\sqrt[3]{4} - 2\right)$
    Absolute maximum: $\left(3, \dfrac{9\sqrt[3]{25} - 8}{4}\right)$

20) Absolute minimum: $\left(4, \dfrac{16 + 3\sqrt[3]{6}}{8}\right)$
    Absolute maximum: $\left(0, \dfrac{16 + 9\sqrt[3]{2}}{8}\right)$

21) Absolute minimum: $\left(\dfrac{\pi}{6}, -4\right)$
    Absolute maximum: $\left(\dfrac{\pi}{4}, -2\sqrt{2}\right)$

22) No absolute minima.
    No absolute maxima.

23) Absolute minimum: $(0, 0)$
    Absolute maximum: $\left(-\dfrac{\pi}{4}, 2\right)$

24) Absolute minimum: $\left(\dfrac{\pi}{2}, 0\right)$
    Absolute maximum: $\left(\dfrac{\pi}{3}, 1\right)$

25) Absolute minimum: $\left(-\dfrac{\pi}{2}, 0\right)$
    Absolute maximum: $\left(-\dfrac{\pi}{3}, \dfrac{1}{2}\right)$

26) Absolute minimum: $(0, 0)$
    Absolute maximum: $\left(-\dfrac{\pi}{2}, 2\right)$

Solutions Practice 29

1) Critical point at: $x = -1$ No discontinuities exist.
   Increasing: $(-1, \infty)$ Decreasing: $(-\infty, -1)$

2) Critical point at: $x = -3$ No discontinuities exist.
   Increasing: $(-3, \infty)$ Decreasing: $(-\infty, -3)$

3) Critical point at: $x = 2$ No discontinuities exist.
   Increasing: $(2, \infty)$ Decreasing: $(-\infty, 2)$

4) Critical point at: $x = 2$ No discontinuities exist.
   Increasing: $(-\infty, 2)$ Decreasing: $(2, \infty)$

5) Critical points at: $x = 0, 2$ No discontinuities exist.
   Increasing: $(-\infty, 0), (2, \infty)$ Decreasing: $(0, 2)$

6) Critical points at: $x = 4, \dfrac{16}{3}$ No discontinuities exist.
   Increasing: $\left(4, \dfrac{16}{3}\right)$ Decreasing: $(-\infty, 4), \left(\dfrac{16}{3}, \infty\right)$

7) Critical points at: $x = 0, \dfrac{4}{3}$ No discontinuities exist.
   Increasing: $\left(0, \dfrac{4}{3}\right)$ Decreasing: $(-\infty, 0), \left(\dfrac{4}{3}, \infty\right)$

8) Critical points at: $x = 0, \dfrac{2}{3}$ No discontinuities exist.
   Increasing: $\left(0, \dfrac{2}{3}\right)$ Decreasing: $(-\infty, 0), \left(\dfrac{2}{3}, \infty\right)$

9) Critical points at: $x = -\dfrac{\sqrt{6}}{2}, 0, \dfrac{\sqrt{6}}{2}$ No discontinuities exist.
   Increasing: $\left(-\dfrac{\sqrt{6}}{2}, 0\right), \left(\dfrac{\sqrt{6}}{2}, \infty\right)$ Decreasing: $\left(-\infty, -\dfrac{\sqrt{6}}{2}\right), \left(0, \dfrac{\sqrt{6}}{2}\right)$

10) Critical points at: $x = -1, 0, 1$ No discontinuities exist.
    Increasing: $(-1, 0), (1, \infty)$ Decreasing: $(-\infty, -1), (0, 1)$

11) Critical points at: $x = -\dfrac{\sqrt{2}}{2}, 0, \dfrac{\sqrt{2}}{2}$ No discontinuities exist.
    Increasing: $\left(-\dfrac{\sqrt{2}}{2}, 0\right), \left(\dfrac{\sqrt{2}}{2}, \infty\right)$ Decreasing: $\left(-\infty, -\dfrac{\sqrt{2}}{2}\right), \left(0, \dfrac{\sqrt{2}}{2}\right)$

12) Critical points at: $x = -\dfrac{\sqrt{6}}{2}, 0, \dfrac{\sqrt{6}}{2}$ No discontinuities exist.
    Increasing: $\left(-\infty, -\dfrac{\sqrt{6}}{2}\right), \left(0, \dfrac{\sqrt{6}}{2}\right)$ Decreasing: $\left(-\dfrac{\sqrt{6}}{2}, 0\right), \left(\dfrac{\sqrt{6}}{2}, \infty\right)$

13) Critical points at: $x = -\dfrac{\sqrt{30}}{5}, 0, \dfrac{\sqrt{30}}{5}$ No discontinuities exist.
    Increasing: $\left(-\dfrac{\sqrt{30}}{5}, \dfrac{\sqrt{30}}{5}\right)$ Decreasing: $\left(-\infty, -\dfrac{\sqrt{30}}{5}\right), \left(\dfrac{\sqrt{30}}{5}, \infty\right)$

14) Critical points at: $x = -\dfrac{3\sqrt{5}}{5}, 0, \dfrac{3\sqrt{5}}{5}$ No discontinuities exist.
    Increasing: $\left(-\infty, -\dfrac{3\sqrt{5}}{5}\right), \left(\dfrac{3\sqrt{5}}{5}, \infty\right)$ Decreasing: $\left(-\dfrac{3\sqrt{5}}{5}, \dfrac{3\sqrt{5}}{5}\right)$

15) Critical points at: $x = -\dfrac{3\sqrt{5}}{5}, 0, \dfrac{3\sqrt{5}}{5}$ No discontinuities exist.
    Increasing: $\left(-\dfrac{3\sqrt{5}}{5}, \dfrac{3\sqrt{5}}{5}\right)$ Decreasing: $\left(-\infty, -\dfrac{3\sqrt{5}}{5}\right), \left(\dfrac{3\sqrt{5}}{5}, \infty\right)$

16) Critical points at: $x = -\dfrac{3\sqrt{5}}{5}, 0, \dfrac{3\sqrt{5}}{5}$ No discontinuities exist.
    Increasing: $\left(-\dfrac{3\sqrt{5}}{5}, \dfrac{3\sqrt{5}}{5}\right)$ Decreasing: $\left(-\infty, -\dfrac{3\sqrt{5}}{5}\right), \left(\dfrac{3\sqrt{5}}{5}, \infty\right)$

17) Critical points at: $x = -\dfrac{\pi}{2}, \dfrac{\pi}{2}$ No discontinuities exist.
    Increasing: $\left(-\pi, -\dfrac{\pi}{2}\right), \left(\dfrac{\pi}{2}, \pi\right)$ Decreasing: $\left(-\dfrac{\pi}{2}, \dfrac{\pi}{2}\right)$

18) No critical points exist. Discontinuities at: $x = -\pi, -\dfrac{\pi}{2}, 0, \dfrac{\pi}{2}, \pi$
    Increasing: No intervals exist. Decreasing: $\left(-\pi, -\dfrac{\pi}{2}\right), \left(-\dfrac{\pi}{2}, 0\right), \left(0, \dfrac{\pi}{2}\right), \left(\dfrac{\pi}{2}, \pi\right)$

19) Critical points at: $x = -\pi, 0, \pi$ No discontinuities exist.
    Increasing: $(0, \pi)$ Decreasing: $(-\pi, 0)$

20) No critical points exist. Discontinuities at: $x = -\frac{\pi}{2}, \frac{\pi}{2}$
    Increasing: No intervals exist. Decreasing: $\left(-\pi, -\frac{\pi}{2}\right), \left(-\frac{\pi}{2}, \frac{\pi}{2}\right), \left(\frac{\pi}{2}, \pi\right)$
21) No critical points exist. Discontinuity at: $x = 3$
    Increasing: $(-\infty, 3), (3, \infty)$ Decreasing: No intervals exist.
22) No critical points exist. Discontinuity at: $x = -3$
    Increasing: $(-\infty, -3), (-3, \infty)$ Decreasing: No intervals exist.
23) No critical points exist. Discontinuity at: $x = -3$
    Increasing: $(-\infty, -3), (-3, \infty)$ Decreasing: No intervals exist.
24) Critical points at: $x = 0, 4$ Discontinuity at: $x = 2$
    Increasing: $(-\infty, 0), (4, \infty)$ Decreasing: $(0, 2), (2, 4)$
25) No critical points exist. Discontinuities at: $x = -1, 1$
    Increasing: $(-\infty, -1), (-1, 1), (1, \infty)$ Decreasing: No intervals exist.
26) Critical points at: $x = -\sqrt{3}, \sqrt{3}$ Discontinuity at: $x = 0$
    Increasing: $(-\infty, -\sqrt{3}), (\sqrt{3}, \infty)$ Decreasing: $(-\sqrt{3}, 0), (0, \sqrt{3})$
27) Critical point at: $x = 0$ Discontinuities at: $x = 4, -4$
    Increasing: $(0, 4), (4, \infty)$ Decreasing: $(-\infty, -4), (-4, 0)$
28) Critical points at: $x = -4, 4$ No discontinuities exist.
    Increasing: $(-\infty, -4), (4, \infty)$ Decreasing: $(-4, 4)$
29) Critical point at: $x = 5$ No discontinuities exist.
    Increasing: No intervals exist. Decreasing: $(-\infty, \infty)$
30) Critical point at: $x = -6$ No discontinuities exist.
    Increasing: $(-6, \infty)$ Decreasing: $(-\infty, -6)$
31) Critical points at: $x = 0, 2$ No discontinuities exist.
    Increasing: $(-\infty, 0)$ Decreasing: $(0, \infty)$
32) Critical points at: $x = -2, 0$ No discontinuities exist.
    Increasing: $(0, \infty)$ Decreasing: $(-\infty, 0)$

Solutions Practice 30

1)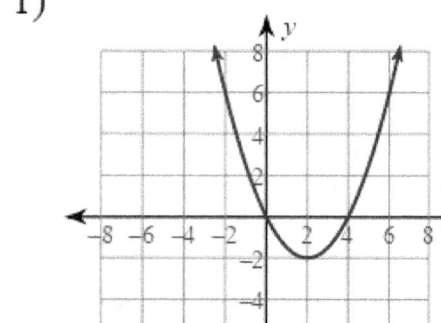
x-intercepts at $x = 0, 4$ y-intercept at $y = 0$
No vertical asymptotes exist.
No horizontal asymptotes exist.
Critical point at: $x = 2$
Increasing: $(2, \infty)$ Decreasing: $(-\infty, 2)$
No inflection points exist.
Concave up: $(-\infty, \infty)$ Concave down: No intervals exist.
Relative minimum: $(2, -2)$ No relative maxima.

2)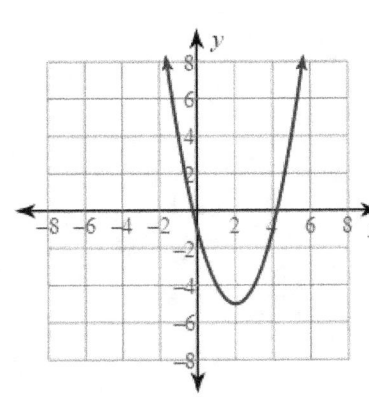
x-intercepts at $x = 2 - \sqrt{5}, 2 + \sqrt{5}$ y-intercept at $y = -1$
No vertical asymptotes exist.
No horizontal asymptotes exist.
Critical point at: $x = 2$
Increasing: $(2, \infty)$ Decreasing: $(-\infty, 2)$
No inflection points exist.
Concave up: $(-\infty, \infty)$ Concave down: No intervals exist.
Relative minimum: $(2, -5)$ No relative maxima.

3)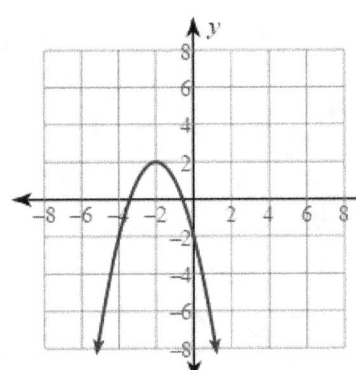
x-intercepts at $x = -2 - \sqrt{2}, -2 + \sqrt{2}$ y-intercept at $y = -2$
No vertical asymptotes exist.
No horizontal asymptotes exist.
Critical point at: $x = -2$
Increasing: $(-\infty, -2)$ Decreasing: $(-2, \infty)$
No inflection points exist.
Concave up: No intervals exist.   Concave down: $(-\infty, \infty)$
No relative minima.   Relative maximum: $(-2, 2)$

4)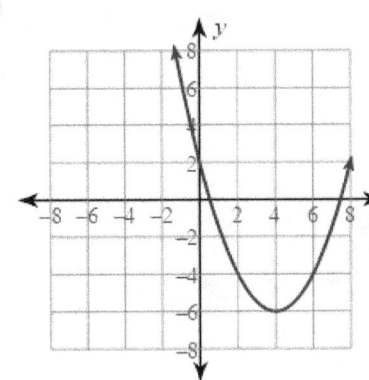
x-intercepts at $x = 4 - 2\sqrt{3}, 4 + 2\sqrt{3}$ y-intercept at $y = 2$
No vertical asymptotes exist.
No horizontal asymptotes exist.
Critical point at: $x = 4$
Increasing: $(4, \infty)$ Decreasing: $(-\infty, 4)$
No inflection points exist.
Concave up: $(-\infty, \infty)$ Concave down: No intervals exist.
Relative minimum: $(4, -6)$ No relative maxima.

5)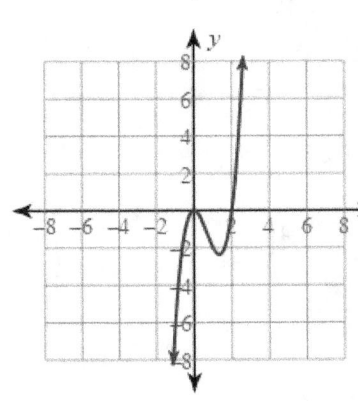
x-intercepts at $x = 0, 2$ y-intercept at $y = 0$
No vertical asymptotes exist.
No horizontal asymptotes exist.
Critical points at: $x = 0, \dfrac{4}{3}$
Increasing: $(-\infty, 0), \left(\dfrac{4}{3}, \infty\right)$ Decreasing: $\left(0, \dfrac{4}{3}\right)$
Inflection point at: $x = \dfrac{2}{3}$
Concave up: $\left(\dfrac{2}{3}, \infty\right)$ Concave down: $\left(-\infty, \dfrac{2}{3}\right)$
Relative minimum: $\left(\dfrac{4}{3}, -\dfrac{64}{27}\right)$ Relative maximum: $(0, 0)$

6)

x-intercepts at $x = -1 - \sqrt{5}, 0, -1 + \sqrt{5}$ y-intercept at $y = 0$
No vertical asymptotes exist.
No horizontal asymptotes exist.
Critical points at: $x = -2, \dfrac{2}{3}$
Increasing: $(-\infty, -2), \left(\dfrac{2}{3}, \infty\right)$ Decreasing: $\left(-2, \dfrac{2}{3}\right)$
Inflection point at: $x = -\dfrac{2}{3}$
Concave up: $\left(-\dfrac{2}{3}, \infty\right)$ Concave down: $\left(-\infty, -\dfrac{2}{3}\right)$
Relative minimum: $\left(\dfrac{2}{3}, -\dfrac{20}{81}\right)$ Relative maximum: $\left(-2, \dfrac{4}{3}\right)$

7)

x-intercepts at $x = -1, 1$ y-intercept at $y = -\dfrac{1}{8}$
No vertical asymptotes exist.
No horizontal asymptotes exist.
Critical points at: $x = -1, 0, 1$
Increasing: $(-\infty, -1), (0, 1)$ Decreasing: $(-1, 0), (1, \infty)$
Inflection points at: $x = -\dfrac{\sqrt{3}}{3}, \dfrac{\sqrt{3}}{3}$
Concave up: $\left(-\dfrac{\sqrt{3}}{3}, \dfrac{\sqrt{3}}{3}\right)$ Concave down: $\left(-\infty, -\dfrac{\sqrt{3}}{3}\right), \left(\dfrac{\sqrt{3}}{3}, \infty\right)$
Relative minimum: $\left(0, -\dfrac{1}{8}\right)$ Relative maxima: $(-1, 0), (1, 0)$

8)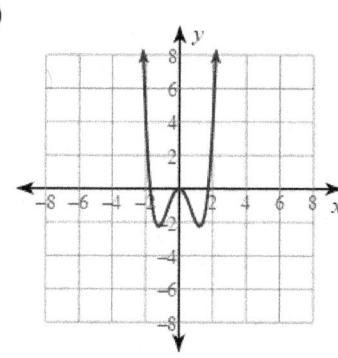

x-intercepts at $x = -\sqrt{3}, 0, \sqrt{3}$ y-intercept at $y = 0$
No vertical asymptotes exist.
No horizontal asymptotes exist.
Critical points at: $x = -\dfrac{\sqrt{6}}{2}, 0, \dfrac{\sqrt{6}}{2}$
Increasing: $\left(-\dfrac{\sqrt{6}}{2}, 0\right), \left(\dfrac{\sqrt{6}}{2}, \infty\right)$ Decreasing: $\left(-\infty, -\dfrac{\sqrt{6}}{2}\right), \left(0, \dfrac{\sqrt{6}}{2}\right)$
Inflection points at: $x = -\dfrac{\sqrt{2}}{2}, \dfrac{\sqrt{2}}{2}$
Concave up: $\left(-\infty, -\dfrac{\sqrt{2}}{2}\right), \left(\dfrac{\sqrt{2}}{2}, \infty\right)$ Concave down: $\left(-\dfrac{\sqrt{2}}{2}, \dfrac{\sqrt{2}}{2}\right)$
Relative minima: $\left(-\dfrac{\sqrt{6}}{2}, -\dfrac{9}{4}\right), \left(\dfrac{\sqrt{6}}{2}, -\dfrac{9}{4}\right)$ Relative maximum: $(0, 0)$

9)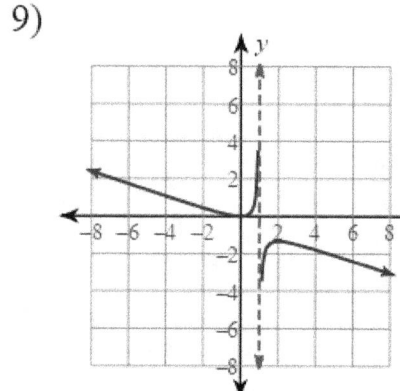

x-intercept at $x = 0$ y-intercept at $y = 0$
Vertical asymptote at: $x = 1$
No horizontal asymptotes exist.
Slant asymptote: $y = -\dfrac{x}{3} - \dfrac{1}{3}$
Critical points at: $x = 0, 2$
Increasing: $(0, 1), (1, 2)$ Decreasing: $(-\infty, 0), (2, \infty)$
No inflection points exist.
Concave up: $(-\infty, 1)$ Concave down: $(1, \infty)$
Relative minimum: $(0, 0)$ Relative maximum: $\left(2, -\dfrac{4}{3}\right)$

10)

x-intercept at $x = 0$ y-intercept at $y = 0$
Vertical asymptote at: $x = 3$
Horizontal asymptote at $y = 3$
No critical points exist.
Increasing: No intervals exist.   Decreasing: $(-\infty, 3), (3, \infty)$
No inflection points exist.
Concave up: $(3, \infty)$ Concave down: $(-\infty, 3)$
No relative minima.   No relative maxima.

11)

No x-intercepts. y-intercept at $y = -\dfrac{1}{3}$
Vertical asymptote at: $x = -3$
Horizontal asymptote at: $y = 0$
No critical points exist.
Increasing: $(-\infty, -3), (-3, \infty)$ Decreasing: No intervals exist.
No inflection points exist.
Concave up: $(-\infty, -3)$ Concave down: $(-3, \infty)$
No relative minima.   No relative maxima.

12)

x-intercept at $x = 0$ y-intercept at $y = 0$
Vertical asymptote at: $x = -1$
No horizontal asymptotes exist.
Slant asymptote: $y = \dfrac{x}{2} - \dfrac{1}{2}$
Critical points at: $x = -2, 0$
Increasing: $(-\infty, -2), (0, \infty)$ Decreasing: $(-2, -1), (-1, 0)$
No inflection points exist.
Concave up: $(-1, \infty)$ Concave down: $(-\infty, -1)$
Relative minimum: $(0, 0)$ Relative maximum: $(-2, -2)$

13)

x-intercepts at $x = -1, 1$ No y-intercepts.
Vertical asymptote at: $x = 0$
Horizontal asymptote at $y = 0$
Critical points at: $x = -\sqrt{3}, \sqrt{3}$
Increasing: $(-\sqrt{3}, 0), (0, \sqrt{3})$ Decreasing: $(-\infty, -\sqrt{3}), (\sqrt{3}, \infty)$
Inflection points at: $x = -\sqrt{6}, \sqrt{6}$
Concave up: $(-\sqrt{6}, 0), (\sqrt{6}, \infty)$ Concave down: $(-\infty, -\sqrt{6}), (0, \sqrt{6})$
Relative minimum: $\left(-\sqrt{3}, -\dfrac{2\sqrt{3}}{9}\right)$ Relative maximum: $\left(\sqrt{3}, \dfrac{2\sqrt{3}}{9}\right)$

14) No x-intercepts. y-intercept at $y = -\dfrac{1}{16}$
Vertical asymptotes at $x = -4, 4$
Horizontal asymptote at $y = 0$
Critical point at: $x = 0$
Increasing: $(-\infty, -4), (-4, 0)$ Decreasing: $(0, 4), (4, \infty)$
No inflection points exist.
Concave up: $(-\infty, -4), (4, \infty)$ Concave down: $(-4, 4)$
No relative minima.   Relative maximum: $\left(0, -\dfrac{1}{16}\right)$

15) No x-intercepts. y-intercept at $y = 1$
No vertical asymptotes exist.
Horizontal asymptote at $y = 0$
Critical point at: $x = 0$
Increasing: $(-\infty, 0)$ Decreasing: $(0, \infty)$
Inflection points at: $x = -1, 1$
Concave up: $(-\infty, -1), (1, \infty)$ Concave down: $(-1, 1)$
No relative minima.   Relative maximum: $(0, 1)$

16) x-intercept at $x = 0$ y-intercept at $y = 0$
Vertical asymptotes at $x = -2, 2$
No horizontal asymptotes exist.
Slant asymptote: $y = x$
Critical points at: $x = -2\sqrt{3}, 0, 2\sqrt{3}$
Increasing: $(-\infty, -2\sqrt{3}), (2\sqrt{3}, \infty)$ Decreasing: $(-2\sqrt{3}, -2), (-2, 2), (2, 2\sqrt{3})$
Inflection point at: $x = 0$
Concave up: $(-2, 0), (2, \infty)$ Concave down: $(-\infty, -2), (0, 2)$
Relative minimum: $(2\sqrt{3}, 3\sqrt{3})$ Relative maximum: $(-2\sqrt{3}, -3\sqrt{3})$

17) x-intercept at $x = 6$ y-intercept at $y = 3\sqrt[3]{12}$
No vertical asymptotes exist.
No horizontal asymptotes exist.
Critical point at: $x = 6$
Increasing: $(6, \infty)$ Decreasing: $(-\infty, 6)$
No inflection points exist.
Concave up: No intervals exist.   Concave down: $(-\infty, 6), (6, \infty)$
Relative minimum: $(6, 0)$ No relative maxima.

18)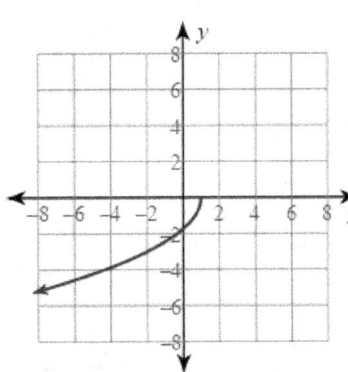
x-intercept at $x = 1$ y-intercept at $y = -\sqrt{3}$
No vertical asymptotes exist.
No horizontal asymptotes exist.
Critical point at: $x = 1$
Increasing: $(-\infty, 1)$ Decreasing: No intervals exist.
No inflection points exist.
Concave up: $(-\infty, 1)$ Concave down: No intervals exist.
No relative minima.   No relative maxima.

19)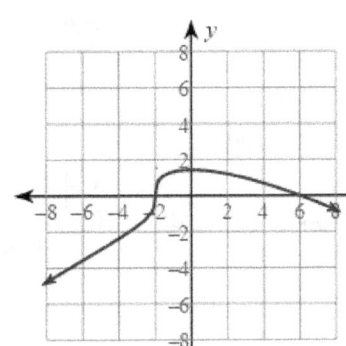
x-intercepts at $x = -2, 6$ y-intercept at $y = \dfrac{9\sqrt[3]{2}}{8}$
No vertical asymptotes exist.
No horizontal asymptotes exist.
Critical points at: $x = -2, 0$
Increasing: $(-\infty, 0)$ Decreasing: $(0, \infty)$
Inflection points at: $x = -6, -2$
Concave up: $(-6, -2)$ Concave down: $(-\infty, -6), (-2, \infty)$
No relative minima.   Relative maximum: $\left(0, \dfrac{9\sqrt[3]{2}}{8}\right)$

20)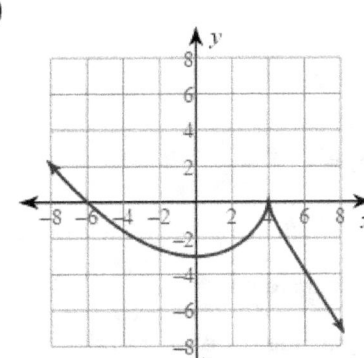
x-intercepts at $x = -6, 4$ y-intercept at $y = -\dfrac{12\sqrt[3]{2}}{5}$
No vertical asymptotes exist.
No horizontal asymptotes exist.
Critical points at: $x = 0, 4$
Increasing: $(0, 4)$ Decreasing: $(-\infty, 0), (4, \infty)$
Inflection point at: $x = 6$
Concave up: $(-\infty, 4), (4, 6)$ Concave down: $(6, \infty)$
Relative minimum: $\left(0, -\dfrac{12\sqrt[3]{2}}{5}\right)$ Relative maximum: $(4, 0)$

Solutions Practice 31

1)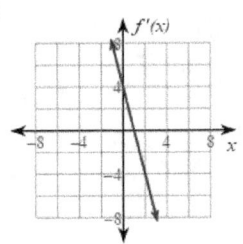

2)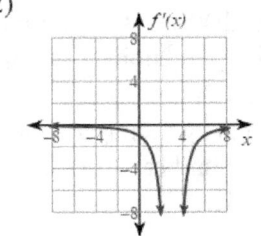

3)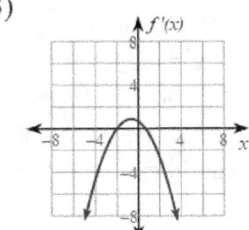

4)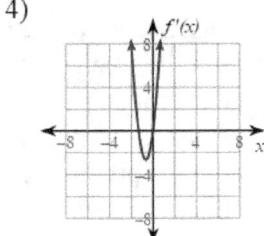

5)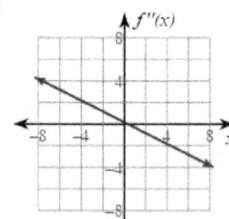
6)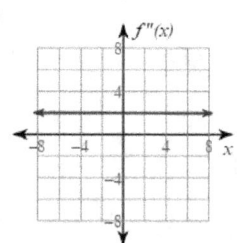
7)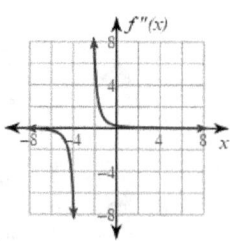
8)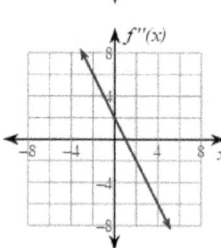

9)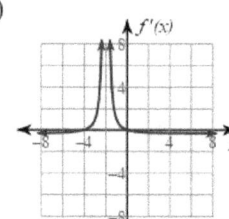
10)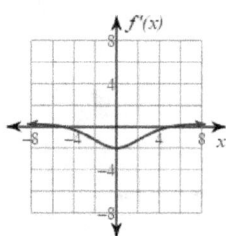
11)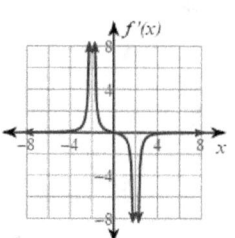
12)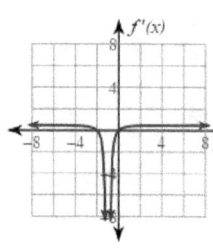

13)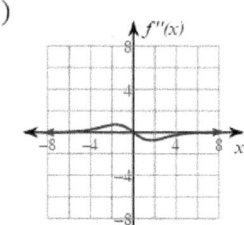
14)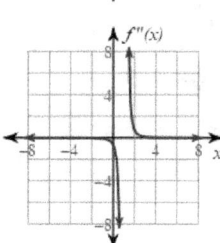
15)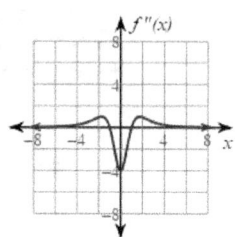
16)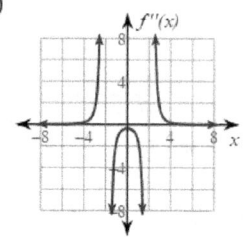

Solutions Practice 32

1) A = the total area of the two corrals x = the length of the non-adjacent sides of each corral
Function to maximize: $A = 2x \cdot \dfrac{200 - 4x}{3}$ where $0 < x < 50$

Dimensions of each corall: 25 ft (non-adjacent sides) by $\dfrac{100}{3}$ ft (adjacent sides)

2) A = the total area of the two corrals x = the length of the non-adjacent sides of each corral
Function to maximize: $A = 2x \cdot \dfrac{400 - 4x}{3}$ where $0 < x < 100$

Dimensions of each corall: 50 ft (non-adjacent sides) by $\dfrac{200}{3}$ ft (adjacent sides)

3) A = the area of the pigpen x = the length of the sides perpendicular to the stone wall
Function to maximize: $A = x(400 - 2x)$ where $0 < x < 200$
Dimensions of the pigpen: 100 ft (perpendicular to wall) by 200 ft (parallel to wall)

4) A = the area of the pigpen x = the length of the sides perpendicular to the stone wall
Function to maximize: $A = x(100 - 2x)$ where $0 < x < 50$
Dimensions of the pigpen: 25 ft (perpendicular to wall) by 50 ft (parallel to wall)

5) P = the product of the two numbers x = the positive number
Function to minimize: $P = x(x - 5)$ where $-\infty < x < \infty$
Smallest product of the two numbers: $-\dfrac{25}{4}$

6) P = the product of the two numbers x = the positive number
Function to minimize: $P = x(x - 8)$ where $-\infty < x < \infty$
Smallest product of the two numbers: -16

7) p = the profit per day x = the number of items manufactured per day
Function to maximize: $p = x(130 - 0.1x) - (60x + 4000)$ where $0 \le x < \infty$
Optimal number of smartphones to manufacture per day: 350

8) p = the profit per day x = the number of items manufactured per day
Function to maximize: $p = x(140 - 0.05x) - (70x + 7000)$ where $0 \le x < \infty$
Optimal number of smartphones to manufacture per day: 700

9) V = the volume of the box x = the length of the sides of the squares
Function to maximize: $V = (16 - 2x)(10 - 2x) \cdot x$ where $0 < x < 5$
Sides of the squares: 2 in

10) V = the volume of the box x = the length of the sides of the squares
Function to maximize: $V = (30 - 2x)(14 - 2x) \cdot x$ where $0 < x < 7$
Sides of the squares: 3 in

11) A = the area of the poster x = the width of the photo
Function to minimize: $A = (x + 2 \cdot 2)\left(\dfrac{50}{x} + 2 \cdot 4\right)$ where $0 < x < \infty$
Dimensions of the entire poster: 9 in wide by 18 in tall

12) A = the area of the poster x = the width of the photo
Function to minimize: $A = (x + 2 \cdot 1)\left(\dfrac{98}{x} + 2 \cdot 2\right)$ where $0 < x < \infty$
Dimensions of the entire poster: 9 in wide by 18 in tall

13) A = the area of the glass x = the length of the sides of the square bottom
Function to minimize: $A = x^2 + 4x \cdot \dfrac{1372}{x^2}$ where $0 < x < \infty$
Dimensions of the aquarium: 14 ft by 14 ft by 7 ft tall

14) A = the area of the glass x = the length of the sides of the square bottom
Function to minimize: $A = x^2 + 4x \cdot \dfrac{2048}{x^2}$ where $0 < x < \infty$
Dimensions of the aquarium: 16 ft by 16 ft by 8 ft tall

15) A = the area of the rectangle x = half the base of the rectangle
Function to maximize: $A = 2x \cdot 2 \cdot \dfrac{\sqrt{49 - x^2}}{2}$ where $0 < x < 7$
Area of largest rectangle: 49

16) A = the area of the rectangle x = half the base of the rectangle
Function to maximize: $A = 2x \cdot 2 \cdot \dfrac{\sqrt{36 - x^2}}{2}$ where $0 < x < 6$
Area of largest rectangle: 36

17) A = the area of the rectangle x = half the base of the rectangle
Function to maximize: $A = 2x\sqrt{3^2 - x^2}$ where $0 < x < 3$
Area of largest rectangle: 9

18) A = the area of the rectangle x = half the base of the rectangle
Function to maximize: $A = 2x\sqrt{8^2 - x^2}$ where $0 < x < 8$
Area of largest rectangle: 64

19) d = the distance from point $(0, 1)$ to a point on the parabola x = the x-coord. of a point on the parabola
Function to minimize: $d = \sqrt{x^2 + (5 - x^2 - 1)^2}$ where $-\infty < x < \infty$
Points on the parabola that are closest to the point $(0, 1)$: $\left(-\dfrac{\sqrt{14}}{2}, \dfrac{3}{2}\right), \left(\dfrac{\sqrt{14}}{2}, \dfrac{3}{2}\right)$

20) d = the distance from point $(0, 3)$ to a point on the parabola x = the x-coord. of a point on the parabola
Function to minimize: $d = \sqrt{x^2 + (4 - x^2 - 3)^2}$ where $-\infty < x < \infty$
Points on the parabola that are closest to the point $(0, 3)$: $\left(-\dfrac{\sqrt{2}}{2}, \dfrac{7}{2}\right), \left(\dfrac{\sqrt{2}}{2}, \dfrac{7}{2}\right)$

21) d = the distance from point $(1, 0)$ to a point on the curve x = the x-coordinate of a point on the curve
Function to minimize: $d = \sqrt{(x - 1)^2 + (\sqrt{x})^2}$ where $-\infty < x < \infty$
Point on the curve that is closest to the point $(1, 0)$: $\left(\dfrac{1}{2}, \dfrac{\sqrt{2}}{2}\right)$

22) d = the distance from point $(5, 0)$ to a point on the curve x = the x-coordinate of a point on the curve
Function to minimize: $d = \sqrt{(x - 5)^2 + (\sqrt{x})^2}$ where $-\infty < x < \infty$
Point on the curve that is closest to the point $(5, 0)$: $\left(\dfrac{9}{2}, \dfrac{3\sqrt{2}}{2}\right)$

23) 12 ft from the short pole (or 24 ft from the long pole)
24) 5 ft from the short pole (or 15 ft from the long pole)
25) $\dfrac{24}{4 + \pi}$ ft (width) by $\dfrac{12}{4 + \pi}$ ft (height)   26) $\dfrac{20}{4 + \pi}$ ft (width) by $\dfrac{10}{4 + \pi}$ ft (height)

Solutions Practice 33

1) 30   2) 20   3) 16   4) 12
5) 24.5   6) 40.5   7) 29   8) 38.5
9) –2   10) –2   11) 2   12) 9

Solutions Practice 34

1) 26   2) 36   3) 52   4) 48
5) –3   6) –22   7) 78   8) 48
9) 40   10) 40   11) –25   12) 0

Solutions Practice 35

1) left endpoint | 19 |
   right endpoint | 15 |
   midpoint | 17.5 |
   trapezoidal rule | 17 |
   Actual | 52/3 ≈ 17.3333

2) left endpoint | 21
   right endpoint | 13
   midpoint | 16.5 |
   trapezoidal rule | 17
   Actual | 50/3 ≈ 16.6667

3) left endpoint | 10.4167
   right endpoint | 6.41667
   midpoint | 7.87302
   trapezoidal rule | 8.41667
   Actual | $5\ln 5$ ≈ 8.0472

4) left endpoint | 3.8
   right endpoint | 5.13333
   midpoint | 4.35902
   trapezoidal rule | 4.46667
   Actual | $\ln 81$ ≈ 4.3944

5) RIGHT
   TRAP
   ACTUAL
   LEFT

6) RIGHT
   ACTUAL
   TRAP
   LEFT

7) LEFT
   TRAP
   ACTUAL
   RIGHT

8) LEFT
   ACTUAL
   TRAP
   RIGHT

9) Right Riemann Sum
   Actual Area Under the Curve
   Left Riemann Sum

10) Left Riemann Sum
    Actual Area Under the Curve
    Right Riemann Sum

Solutions Practice 36

1) $\sum_{i=0}^{7}(-|i-4|+4)$

2) $\sum_{i=1}^{8}(-|i-4|+4)$

3) $\sum_{i=1}^{4}(-|2i-4|+4)\cdot 2$

4) $\sum_{i=0}^{3}(-|2i-4|+4)\cdot 2$

5) $\sum_{i=0}^{15}(-|0.5i-4|+4)\cdot 0.5$

6) $\sum_{i=1}^{16}(-|0.5i-4|+4)\cdot 0.5$

7) $\sum_{i=1}^{4}(-|i-1|+6)$

8) $\sum_{i=1}^{4}(-(i-2)^2+4)$

9) $\sum_{i=0}^{3}(-(i-2)^2+4)$

10) $\sum_{i=0}^{3}(-|i-1|+6)$

11) $\sum_{i=1}^{12}(-|0.5i-1|+5)\cdot 0.5$

12) $\sum_{i=0}^{16}(-|0.25i-4|+5)\cdot 0.25$

13) $\lim_{n\to\infty}\sum_{i=1}^{n}\cos\left(\pi\cdot\frac{i}{n}\right)\cdot\frac{\pi}{n}$

14) $\lim_{n\to\infty}\sum_{i=0}^{n-1}\sin\left(\pi\cdot\frac{i}{n}\right)\cdot\frac{\pi}{n}$

15) $\lim_{n\to\infty}\sum_{i=1}^{n}\ln\left(1+\frac{ie-i}{n}\right)\cdot\frac{e-1}{n}$

16) $\lim_{n\to\infty}\sum_{i=0}^{n-1}\sin\left(\frac{\pi}{2}+\frac{i\cdot\pi}{2n}\right)\cdot\frac{\pi}{2n}$

17) $\int_{4}^{8} x^2\, dx$

18) $\int_{9}^{16} \sqrt{x}\, dx$

19) $\int_{1}^{e} \ln x\, dx$

20) $\int_{1}^{5} e^x\, dx$

Solutions Practice 37

1) $\int_8^2 g(x)\,dx = -6$

$\int_2^8 (f(x) + g(x))\,dx = 4$

$\int_8^2 3g(x)\,dx = -18$

$\int_8^8 f(x)\,dx = 0$

$\int_2^8 (f(x) - g(x))\,dx = -8$

$\int_2^4 f(x)\,dx + \int_4^8 f(x)\,dx = -2$

$\int_8^2 (2f(x) - 3g(x))\,dx = 22$

2) $\int_a^b (f(x) + g(x))\,dx = 5x - 2$

$\int_b^a g(x)\,dx = -3x + 6$

$\int_b^a 3g(x)\,dx = -9x + 18$

$\int_a^a f(x)\,dx = 0$

$\int_b^a (f(x) - g(x))\,dx = x - 10$

$\int_b^a (3f(x) - 2g(x))\,dx = -24$

$\int_a^c 2f(x)\,dx + \int_c^b 2f(x)\,dx = 4x + 8$

3) $\int_{-2}^{10} f(x)\,dx = 16$

$\int_{-2}^{10} g(x)\,dx = -2$

$\int_{-2}^{-2} f(x)\,dx = 0$

$\int_{-2}^{6} (f(x) + g(x))\,dx = 8$

$\int_6^{10} 2f(x)\,dx = 28$

$\int_2^2 f(x)\,dx = 0$

4) $\int_a^c g(x)\,dx = -x + 4$

$\int_a^c f(x)\,dx = x - 3$

$\int_b^c 3f(x)\,dx = -3x - 12$

$\int_b^b f(x)\,dx = 0$

$\int_a^b (f(x) + g(x))\,dx = 5x - 1$

$\int_b^b (f(x) + g(x))\,dx = 0$

$$\int_{10}^{6} f(x)\,dx = -14 \qquad\qquad \int_{c}^{b} f(x)\,dx = x+4$$

$$\int_{-2}^{10} (f(x)+g(x))\,dx = 14 \qquad\qquad \int_{b}^{a} (f(x)-g(x))\,dx = x-3$$

$$\int_{6}^{-2} (f(x)-g(x))\,dx = 4 \qquad\qquad \int_{a}^{c} (f(x)+g(x))\,dx = 1$$

$$\int_{10}^{-2} (3f(x)-2g(x))\,dx = -52 \qquad\qquad \int_{c}^{a} (2f(x)-3g(x))\,dx = -x+6$$

Solutions Practice 38

1) $F'(x) = x^2 + 4x - 1$ 2) $F'(x) = -x^2 - 8x - 13$ 3) $F'(x) = x^2 + 6x + 11$
4) $F'(x) = 2x - 1$ 5) 4   6) 1   7) 3
8) 36   9) 64   10) $\dfrac{1}{2}$ 11) $F'(x) = -\sin x$

12) $F'(x) = 2\sec^2 x$ 13) $F'(x) = -4x^3 + 4x$ 14) $F'(x) = -6x^5 + 3x^2$ 15) $F'(x) = 2x^5 + 4x^3$
16) $F'(x) = 8x + 4$ 17) $F'(x) = -3x^{11} + 9x^8 - 6x^2$ 18) $F'(x) = 3x^5 + 3x^2$
19) $F'(x) = -2x\sin x^2$ 20) $F'(x) = 3x^2 \csc x^3 \cot x^3$

Solutions Practice 39

1) $2x^5 - 2x^4 + x + C$ 2) $3x^5 + 3x^3 + 4x^2 + C$ 3) $5x^6 + 5x^3 + 4x^2 + C$ 4) $-2x^6 + 3x^5 - x + C$
5) $2x^4 - 5x^2 + 3x + C$ 6) $5x^5 - x^3 - 4x + C$
7) $-\dfrac{3}{x} - \dfrac{4}{x^3} + \dfrac{1}{x^4} + C$ 8) $-\dfrac{3}{x^2} + \dfrac{3}{x^3} + \dfrac{1}{x^4} + C$

9) $-\dfrac{2}{x^2} + \dfrac{3}{x^3} - \dfrac{3}{x^4} + C$ 10) $\dfrac{4}{x} + \dfrac{4}{x^3} - \dfrac{5}{x^4} + C$ 11) $\dfrac{1}{x} + \dfrac{3}{x^2} + \dfrac{3}{x^4} + C$ 12) $\dfrac{2}{x} + \dfrac{1}{x^2} + \dfrac{5}{x^3} + C$

13) $x^{\frac{5}{2}} + 4x^{\frac{9}{4}} + 3x^{\frac{4}{3}} + C$ 14) $2x^{\frac{5}{3}} + 2x^{\frac{4}{3}} + 4x^{\frac{5}{4}} + C$ 15) $2x^{\frac{5}{3}} + 5x^{\frac{7}{5}} - 5x^{\frac{6}{5}} + C$

16) $5x^{\frac{5}{3}} - 2x^{\frac{7}{5}} + 5x^{\frac{6}{5}} + C$ 17) $2x^{\frac{8}{3}} + x^{\frac{7}{3}} + x^{\frac{6}{5}} + C$ 18) $2x^{\frac{7}{5}} + 5x^{\frac{5}{4}} + x^{\frac{6}{5}} + C$
19) $2x^6 + 5x^3 - 3x^2 + C$ 20) $3x^5 + 4x^4 + 4x^3 + C$ 21) $2x^4 + x^3 + 5x + C$
22) $3x^6 + 3x^4 - 3x^2 + C$
23) $\dfrac{1}{x} + \dfrac{5}{x^3} + \dfrac{1}{x^4} + C$ 24) $\dfrac{2}{x} + \dfrac{2}{x^2} + \dfrac{3}{x^3} + C$

25) $4x^6 + x^4 + 4x^2 + C$
26) $-x^3 + \dfrac{3}{x} - \dfrac{4}{x^2} + C$ 27) $x^5 - 5x^{\frac{5}{3}} + 3x^{\frac{4}{3}} + C$ 28) $4x^{\frac{5}{4}} + \dfrac{4}{x} + \dfrac{2}{x^3} + C$

29) $x^{\frac{7}{4}} - \dfrac{2}{x} + \dfrac{3}{x^3} + C$ 30) $2x^{\frac{6}{5}} + \dfrac{4}{x^2} + \dfrac{3}{x^3} + C$

Solutions Practice 40

1) $\frac{1}{6}(3x^2+1)^6+C$
2) $\frac{1}{5}(2x^4+1)^5+C$
3) $\frac{1}{6}(5x^5-3)^6+C$
4) $\frac{1}{5}(3x^5+5)^5+C$
5) $-\frac{1}{3(3x^3+5)^3}+C$
6) $-\frac{1}{2(2x^3+3)^2}+C$
7) $-\frac{1}{4(4x^5-3)^4}+C$
8) $-\frac{1}{3(2x^2+5)^3}+C$
9) $\frac{2}{5}(3x^3-5)^{\frac{5}{2}}+C$
10) $\frac{3}{5}(2x^2+1)^{\frac{5}{3}}+C$
11) $\frac{3}{4}(2x^2+3)^{\frac{4}{3}}+C$
12) $\frac{3}{4}(3x^4-1)^{\frac{4}{3}}+C$
13) $\frac{5}{4}(5x^4-2)^4+C$
14) $\frac{2}{5}(2x^2+5)^5+C$
15) $3(4x^2+3)^{\frac{4}{3}}+C$
16) $\frac{3}{2}(5x^4+1)^{\frac{4}{3}}+C$
17) $\frac{1}{45}(3x-2)^5+\frac{1}{18}(3x-2)^4+C$
18) $\frac{2}{45}(3x-1)^5+\frac{1}{18}(3x-1)^4+C$
19) $-\frac{5}{3(x-3)^3}-\frac{15}{4(x-3)^4}+C$
20) $\frac{15}{112}(4x+5)^{\frac{7}{3}}-\frac{75}{64}(4x+5)^{\frac{4}{3}}+C$
21) $\frac{2}{5}(e^{3x}-5)^5+C$
22) $(e^{3x}+4)^4+C$
23) $(2+\ln-3x)^4+C$
24) $\frac{1}{3}\cdot\sin^6 5x+C$
25) $\frac{1}{2}\cdot\cos^4-4x+C$
26) $\frac{1}{3}\cdot\sin^6 2x+C$
27) $-\frac{3}{4(e^{5x}+2)^4}+C$
28) $\frac{12}{5}\cdot(\csc 3x)^{\frac{5}{3}}+C$
29) $\frac{4}{3}\cdot(\cot 2x)^{\frac{3}{2}}+C$
30) $-\frac{5}{4(-2+\ln 3x)^4}+C$

Solutions Practice 41

1) $-e^x+C$
2) $-5e^x+C$
3) $4e^x+C$
4) $2e^x+C$
5) $\frac{2\cdot 3^x}{\ln 3}+C$
6) $-\frac{5\cdot 4^x}{\ln 4}+C$
7) $-\frac{2\cdot 2^x}{\ln 2}+C$
8) $-\frac{2\cdot 3^x}{\ln 3}+C$
9) $-4\ln|x|+C$
10) $-\ln|x|+C$
11) $-4\ln|x|+C$
12) $3\ln|x|+C$
13) $5e^{x^4-3}+C$
14) $-3e^{2x^3-5}+C$
15) $-3\ln|x^2-2|+C$
16) $-4\ln|5x^5+2|+C$
17) $3\ln|x^3-4|+C$
18) $-4e^{x^5-3}+C$
19) $\ln|\cos 5x|+C$
20) $3\ln|\cot x|+C$
21) $2\ln|\tan 4x|+C$
22) $-4\ln|\sin 2x|+C$
23) $-5\ln|e^{3x}-2|+C$
24) $-3\ln(e^x+5)+C$
25) $-3\ln(e^{5x}+3)+C$
26) $-4\ln|e^{3x}-3|+C$
27) $-3\ln|-2+\ln 2x|+C$
28) $2\ln|-2+\ln-2x|+C$
29) $-5\ln|5+\ln-x|+C$
30) $3\ln|-3+\ln-3x|+C$

Solutions Practice 42

1) $-4\cos x + C$
2) $3\sin x + C$
3) $5\cos x + C$
4) $-3\sin x + C$
5) $3\tan x + C$
6) $-\cot x + C$
7) $2\cot x + C$
8) $4\tan x + C$
9) $4\csc x + C$
10) $-5\sec x + C$
11) $4\sec x + C$
12) $-2\csc x + C$
13) $-4\cot(x^2 + 1) + C$
14) $2\cos(4x^3 - 1) + C$
15) $-4\csc(2x^5 + 5) + C$
16) $-2\tan(x^2 + 4) + C$
17) $4\sin(3x^4 + 4) + C$
18) $-5\cos(5x^4 + 1) + C$
19) $4\sec(5x^3 - 4) + C$
20) $5\tan(x^2 + 5) + C$
21) $2\cot(\csc 4x) + C$
22) $-3\sin(e^{4x} - 5) + C$
23) $4\cot(-3 + \ln x) + C$
24) $\sin(\tan 3x) + C$
25) $-4\ln|\sin x| + C$
26) $3\ln|\csc x - \cot x| + C$
27) $3\ln|\sec x + \tan x| + C$
28) $-\ln|\sec x| + C$
29) $4\cos x + C$
30) $-4\cot x + C$
31) $3\tan x + C$
32) $-4\sin x + C$
33) $\sec(e^{3x} - 4) + C$
34) $\sin(\cos 2x) + C$
35) $-2\sin(\csc 3x) + C$
36) $-\tan(x^2 + 5) + C$
37) $-5\ln|\sin(\cos -2x)| + C$
38) $4\csc(2x^4 - 5) + C$
39) $2\sec(4 + \ln 4x) + C$
40) $3\ln|\csc(e^{5x} - 1) - \cot(e^{5x} - 1)| + C$

Solutions Practice 43

1) -9
2) $-\dfrac{35}{2} = -17.5$
3) $\dfrac{55}{6} \approx 9.167$
4) $-\dfrac{40}{3} \approx -13.333$
5) $\dfrac{41}{12} \approx 3.417$
6) $-\dfrac{21}{4} = -5.25$
7) 0
8) -4
9) $\dfrac{9\sqrt[3]{6} - 6\sqrt[3]{4}}{4} \approx 1.707$
10) $\dfrac{15\sqrt[3]{2}}{4} \approx 4.725$
11) $\dfrac{3}{32} \approx 0.094$
12) $\dfrac{3}{2} = 1.5$
13) $\ln 5 - \ln 2 \approx 0.916$
14) $3\ln 2 \approx 2.079$
15) $\dfrac{e^3 - 1}{e^2} \approx 2.583$
16) $\dfrac{e^2 - 1}{e^2} \approx 0.865$
17) -2
18) $2\sqrt{2} - 2 \approx 0.828$
19) -1
20) $\dfrac{6 + 2\sqrt{3}}{3} \approx 3.155$

Solutions Practice 44

1) $\sin^{-1}\dfrac{x}{3} + C$
2) $\sin^{-1}\dfrac{x}{5} + C$
3) $\dfrac{1}{3}\cdot\tan^{-1}\dfrac{x}{3} + C$
4) $\tan^{-1} x + C$
5) $\dfrac{1}{2}\cdot\sec^{-1}\dfrac{|x|}{2} + C$
6) $\sec^{-1}|x| + C$
7) $\dfrac{1}{3}\cdot\sec^{-1}\dfrac{|5x^2|}{3} + C$
8) $\dfrac{1}{2}\cdot\tan^{-1}\dfrac{2x^3}{2} + C$
9) $\sin^{-1}\dfrac{4x^2}{5} + C$
10) $\dfrac{1}{4}\cdot\tan^{-1}\dfrac{2x^5}{4} + C$
11) $\dfrac{1}{2}\cdot\sec^{-1}\dfrac{|2x^3|}{2} + C$
12) $\dfrac{1}{4}\cdot\tan^{-1}\dfrac{x^4}{4} + C$
13) $\dfrac{1}{4}\cdot\sec^{-1}\dfrac{|4x^3|}{4} + C$
14) $\sin^{-1}\dfrac{2x^3}{3} + C$
15) $\sin^{-1} e^{3x} + C$
16) $\dfrac{1}{2}\cdot\sec^{-1}\dfrac{|\ln -3x|}{2} + C$
17) $\sin^{-1}\dfrac{\ln x}{3} + C$
18) $\tan^{-1}\ln -2x + C$
19) $\dfrac{1}{4}\cdot\sec^{-1}\dfrac{|\ln -3x|}{4} + C$
20) $\sin^{-1} e^x + C$

APPENDIX I: TABLES

Table 1: Calculator Quick Reference

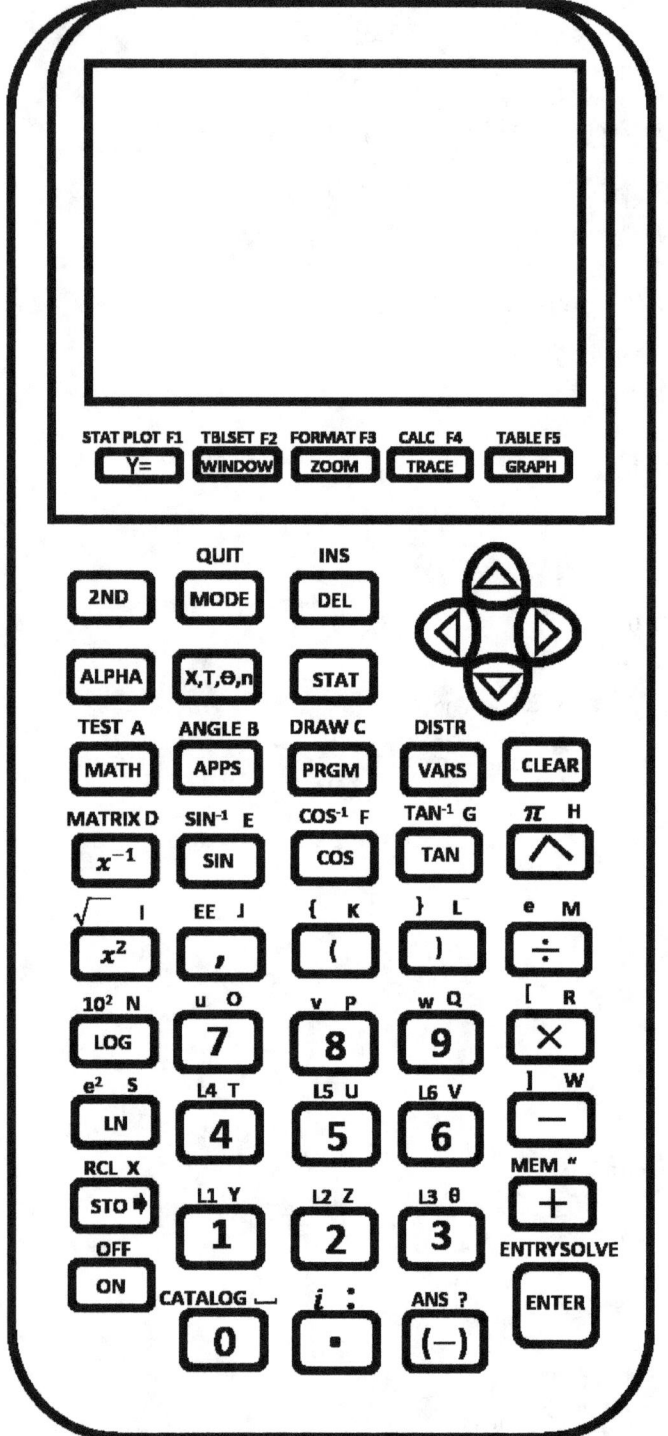

Evaluate the derivative of

$$f(x) = x^2 \text{ at } x = 3$$

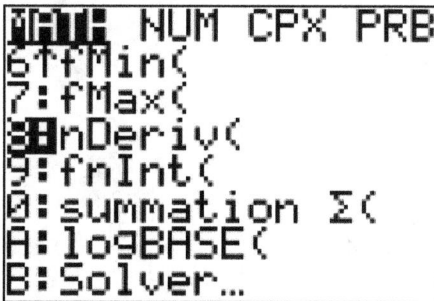

Classic version:

Evaluate the integral

$$\int_1^3 x^2 \, dx$$

TEST A
[MATH]

```
MATH  NUM  CPX  PRB
6↑fMin(
7:fMax(
8:nDeriv(
9:fnInt(
0:summation Σ(
A:logBASE(
B:Solver…
```

$\int_1^3 (X^2) dX$
       8.666666667

Classic version:

fnInt(X²,X,1,3)
       8.666666667

Evaluate the summation

$$\sum_{x=1}^{3} x^2$$

TEST A
[MATH]

```
MATH  NUM  CPX  PRB
6↑fMin(
7:fMax(
8:nDeriv(
9:fnInt(
0:summation Σ(
A:logBASE(
B:Solver…
```

$\sum_{X=1}^{3} (X^2)$
              14

Classic version:

sum(seq(X²,X,1,3)
)
              14

Table 2: Properties of Limits

Properties of Limits
Let b and c be real numbers, let n be a positive integer, and let f and g be functions with the following limits $\lim_{x \to c} f(x) = L$: and $\lim_{x \to c} g(x) = K$
1. Constant: $\lim_{x \to c} b = b$
2. Scalar Multiple: $\lim_{x \to c} [bf(x)] = bL$
3. Sum or Difference: $\lim_{x \to c} [f(x) \pm g(x)] = L \pm K$
4. Product: $\lim_{x \to c} [f(x)g(x)] = LK$
5. Quotient: $\lim_{x \to c} \left[\dfrac{f(x)}{g(x)}\right] = \dfrac{L}{K}$, as long as $K \neq 0$
6. Power: $\lim_{x \to c} [f(x)]^n = L^n$

Table 3: Properties of Derivatives

Properties of Derivatives
1. Constant Rule: $f(x) = c$ then $f'(x) = 0$
2. Constant Multiple Rule: if $g(x) = c \cdot f(x)$ then $g'(x) = c \cdot f'(x)$
3. Power Rule: if $f(x) = x^n$ then $f'(x) = nx^{n-1}$
4. Sum or Difference: if $h(x) = f(x) \pm g(x)$ then $h'(x) = f'(x) \pm g'(x)$
5. Product Rule: if $h(x) = f(x)g(x)$ then $h'(x) = f(x)g'(x) + g(x)f'(x)$
6. Quotient Rule: if $h(x) = \frac{f(x)}{g(x)}$ then $h'(x) = \frac{g(x)f'(x) - f(x)g'(x)}{g(x)^2}$
7. Chain Rule: if $h(x) = f(g(x))$ then $h'(x) = f'(g(x))g'(x)$

Table 4: Derivative Formulas

$\dfrac{d}{dx} au = a \dfrac{du}{dx}$	$\dfrac{d}{dx} u^n = n u^{n-1} \dfrac{du}{dx}$
$\dfrac{d}{dx} e^u = e^u \dfrac{du}{dx}$	$\dfrac{d}{dx} \ln u = \dfrac{1}{u} \dfrac{du}{dx}$
$\dfrac{d}{dx} a^u = a^u \ln a \dfrac{du}{dx}$	$\dfrac{d}{dx} \log_a u = \dfrac{1}{u \ln a} \dfrac{du}{dx}$
$\dfrac{d}{dx} \sin u = \cos u \dfrac{du}{dx}$	$\dfrac{d}{dx} \cos u = -\sin u \dfrac{du}{dx}$
$\dfrac{d}{dx} \tan u = \sec^2 u \dfrac{du}{dx}$	$\dfrac{d}{dx} \cot u = -\csc^2 u \dfrac{du}{dx}$
$\dfrac{d}{dx} \sec u = \sec u \tan u \dfrac{du}{dx}$	$\dfrac{d}{dx} \csc u = -\csc u \cot u \dfrac{du}{dx}$
$\dfrac{d}{dx} \sin^{-1} u = \dfrac{1}{\sqrt{1-u^2}} \dfrac{du}{dx}$	$\dfrac{d}{dx} \cos^{-1} u = -\dfrac{1}{\sqrt{1-u^2}} \dfrac{du}{dx}$
$\dfrac{d}{dx} \tan^{-1} u = \dfrac{1}{1+u^2} \dfrac{du}{dx}$	$\dfrac{d}{dx} \cot^{-1} u = -\dfrac{1}{1+u^2} \dfrac{du}{dx}$
$\dfrac{d}{dx} \sec^{-1} u = \dfrac{1}{|u|\sqrt{u^2-1}} \dfrac{du}{dx}$	$\dfrac{d}{dx} \csc^{-1} u = -\dfrac{1}{|u|\sqrt{u^2-1}} \dfrac{du}{dx}$

Table 5: Common Formulas for Related Rates

Circles	Triangles
$A = \pi r^2$	$A = \dfrac{1}{2}bh$
$C = 2\pi r$	$a^2 + b^2 = c^2$

Spheres	Cones
$V = \dfrac{4}{3}\pi r^3$	$V = \dfrac{1}{3}\pi r^2 h$
$SA = 4\pi r^2$	$SA = \pi r\sqrt{r^2 + h^2} + \pi r^2$

Cylinders	Rectangular Prism
$V = \pi r^2 h$	$V = lwh$
$SA = 2\pi rh + 2\pi r^2$	$SA = 2lw + 2lh + 2wh$

Table 6: Properties of Definite Integrals

Properties of Definite Integrals
1. $\int_a^a f(x)dx = 0$ The integral at a single point is zero. There is no area for a rectangle of width zero.
2. $\int_a^b f(x)dx = -\int_b^a f(x)dx$ The accumulation from a to b is the opposite of going backwards from b to a.
3. $\int_a^b k \cdot f(x)dx = k\int_a^b f(x)dx$ A constant can be factored in/out of an integral.
4. $\int_a^b (f(x) + g(x))dx = \int_a^b f(x)dx + \int_a^b g(x)dx$ Add integrals
5. $\int_a^b (f(x) - g(x))dx = \int_a^b f(x)dx - \int_a^b g(x)dx$ Subtract integrals
6. $\int_a^c f(x)dx + \int_c^b f(x)dx = \int_a^b f(x)dx$ We can add up adjacent integrals to form one integral

Table 7: Integral Formulas

$\int du = u + C$	$\int u^n\, du = \dfrac{u^{n+1}}{n+1} + C,\ n \neq -1$
$\int e^u\, du = e^u + C$	$\int a^u\, du = \dfrac{a^u}{\ln a} + C$
$\int \dfrac{1}{u}\, du = \ln|x| + C$	$\int \cos u\, du = \sin u + C$
$\int \sin u\, du = -\cos u + C$	$\int \sec^2 u\, du = \tan u + C$
$\int \csc^2 u\, du = -\cot u + C$	$\int \sec u \tan u\, du = \sec u + C$
$\int \csc u \cot u\, du = -\csc u + C$	$\int \dfrac{1}{\sqrt{a^2 - u^2}}\, du = \sin^{-1}\dfrac{u}{a} + C$
$\int \dfrac{1}{u^2 + a^2}\, du = \dfrac{1}{a}\tan^{-1}\dfrac{u}{a} + C$	$\int \dfrac{1}{u\sqrt{u^2 - a^2}}\, du = \dfrac{1}{a}\sec^{-1}\dfrac{|u|}{a} + C$

References

Khan Academy. (2020). *Limit of sin(x)/x as x approaches 0*. Retrieved 16 2020, March, from Khan Academy: https://www.khanacademy.org/math/ap-calculus-ab/ab-limits-new/ab-1-8/v/sinx-over-x-as-x-approaches-0

Khan Academy. (2020). *Strategies in Finding Limits*. Retrieved March 14, 2020, from Khan Academy: https://www.khanacademy.org/math/ap-calculus-ab/ab-limits-new/ab-1-7/a/limit-strategies-flow-chart

Wikipedia contributors. (2020, March 2). *Floor and ceiling functions*. Retrieved March 14, 2020, from Wikipedia, The Free Encyclopedia.: https://en.wikipedia.org/wiki/Floor_and_ceiling_functions

GRAPH PAPER

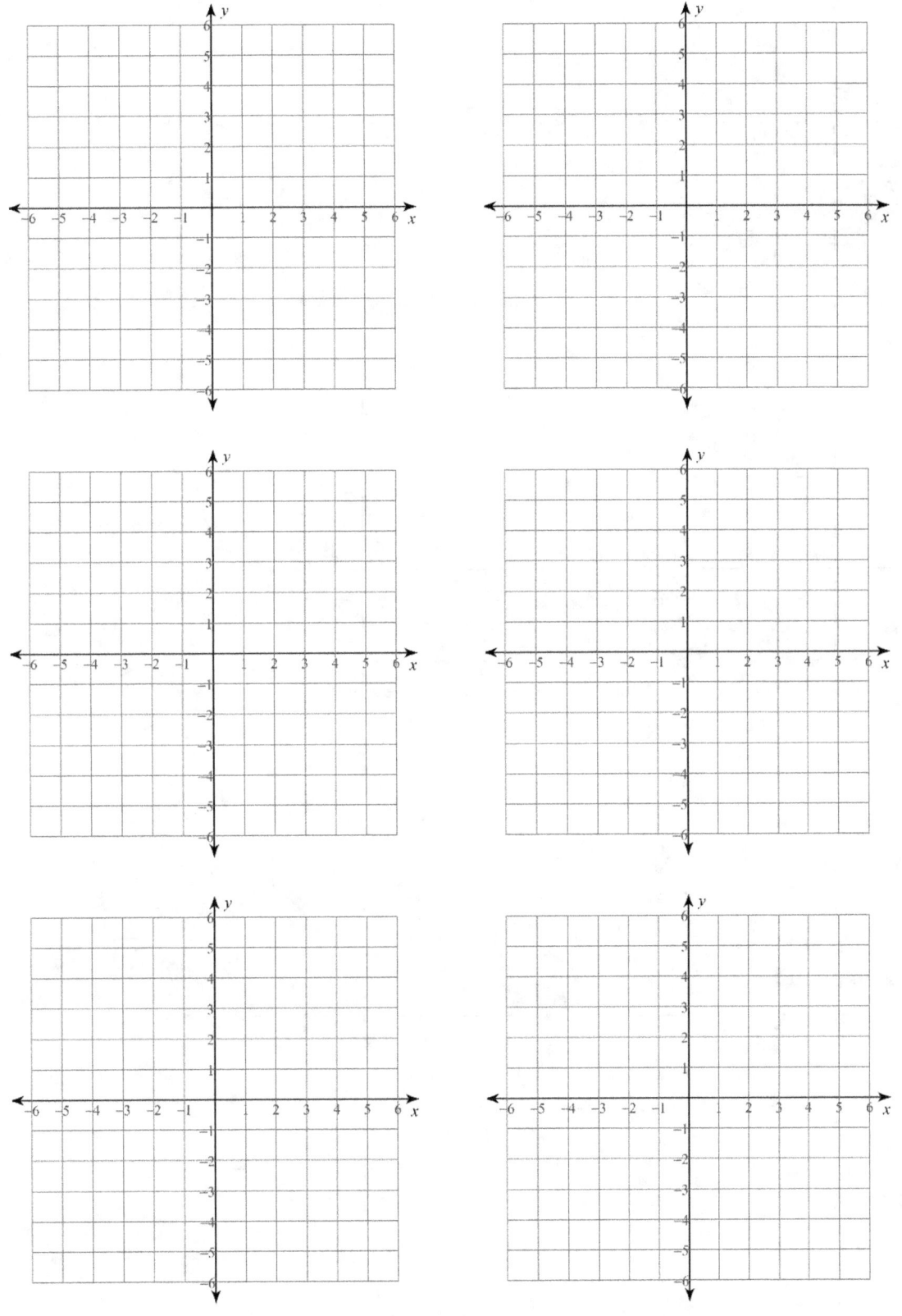

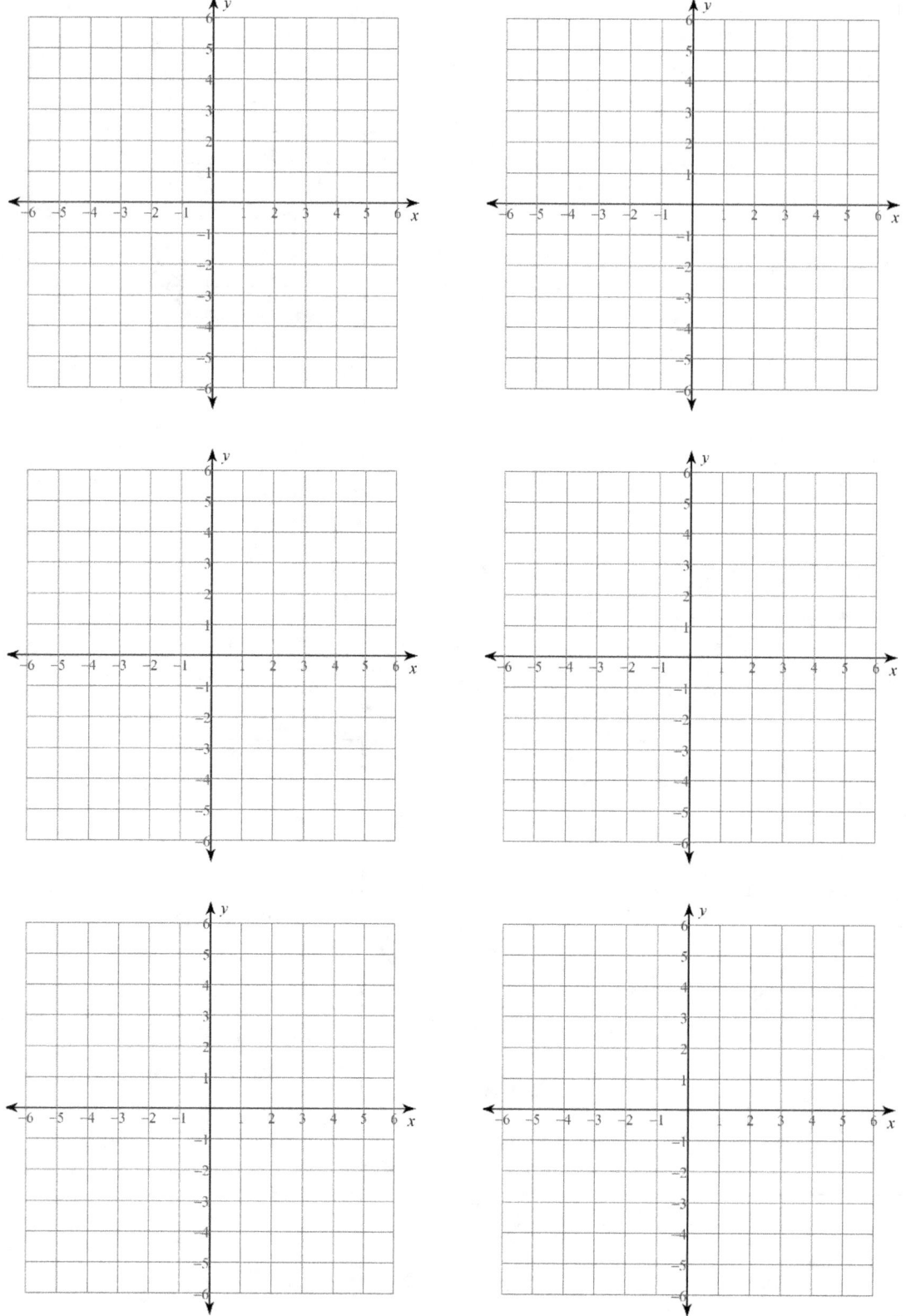

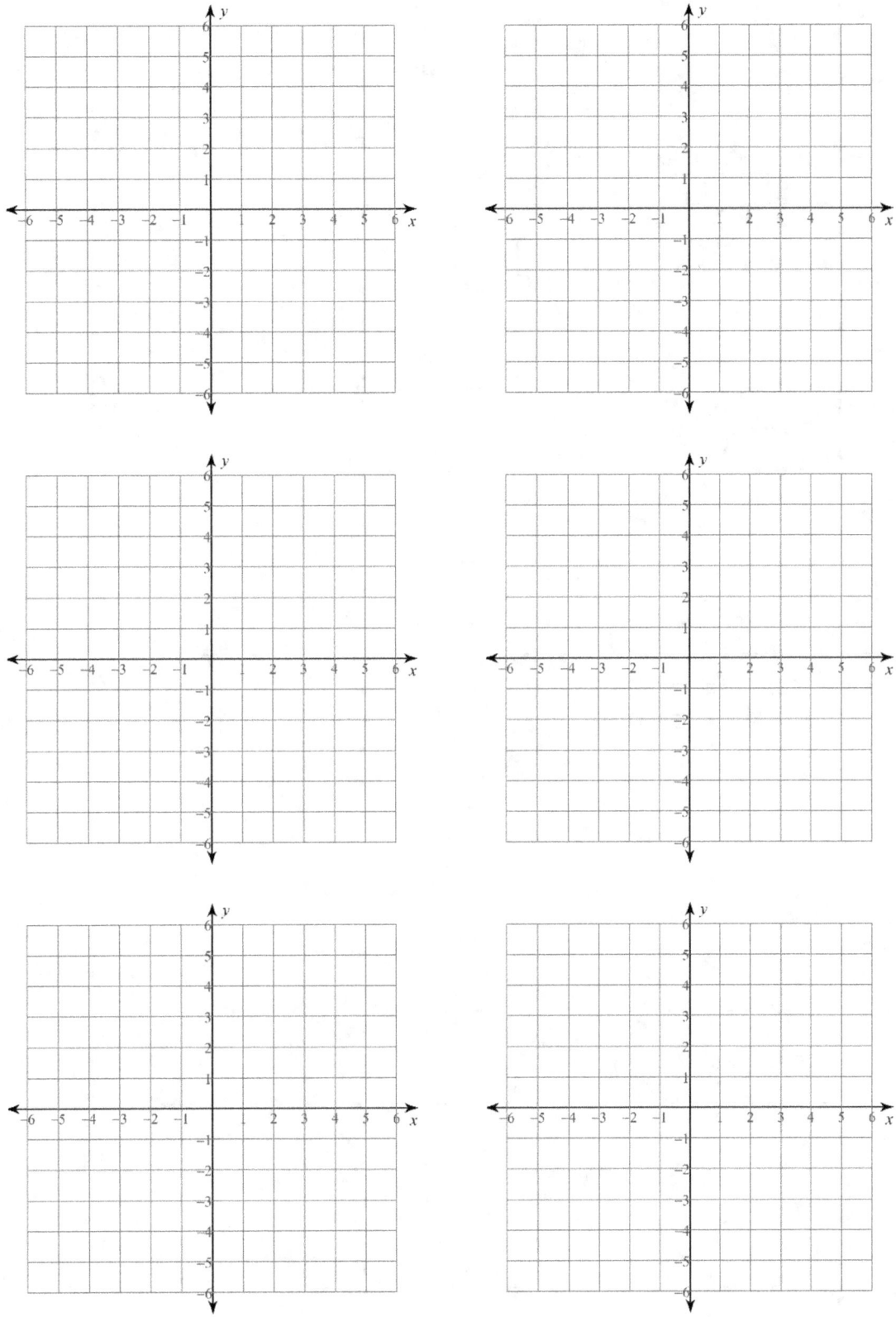

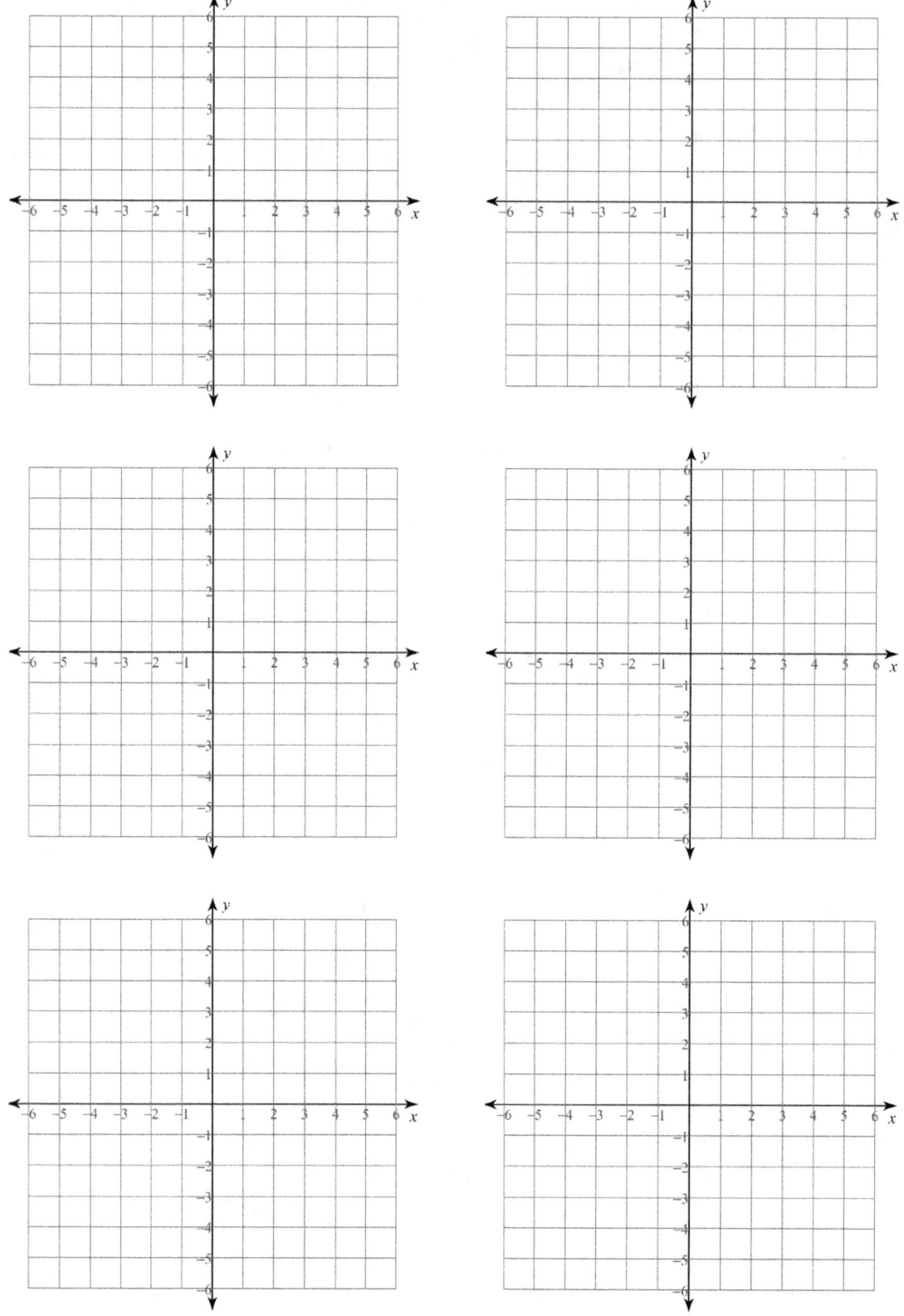

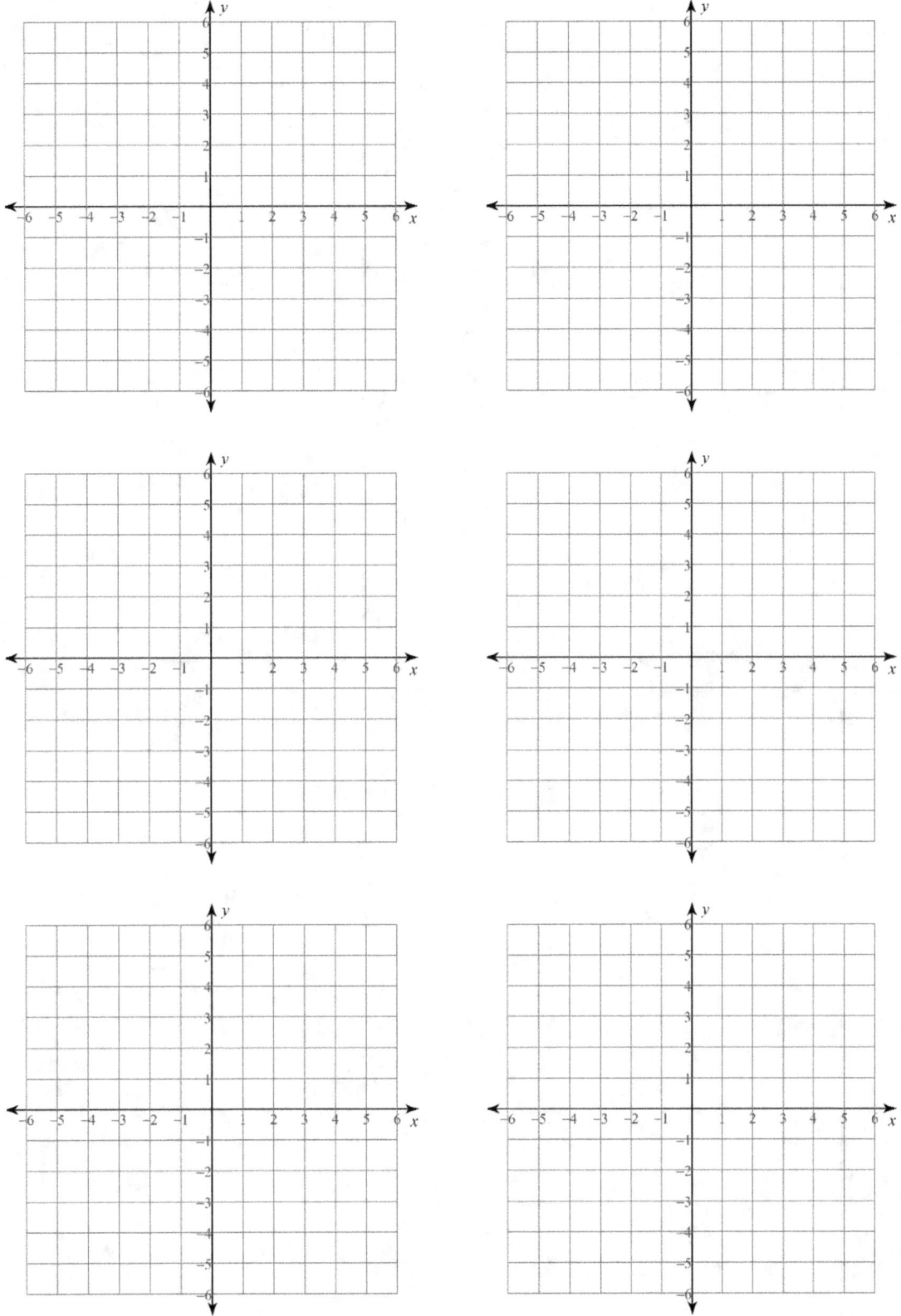

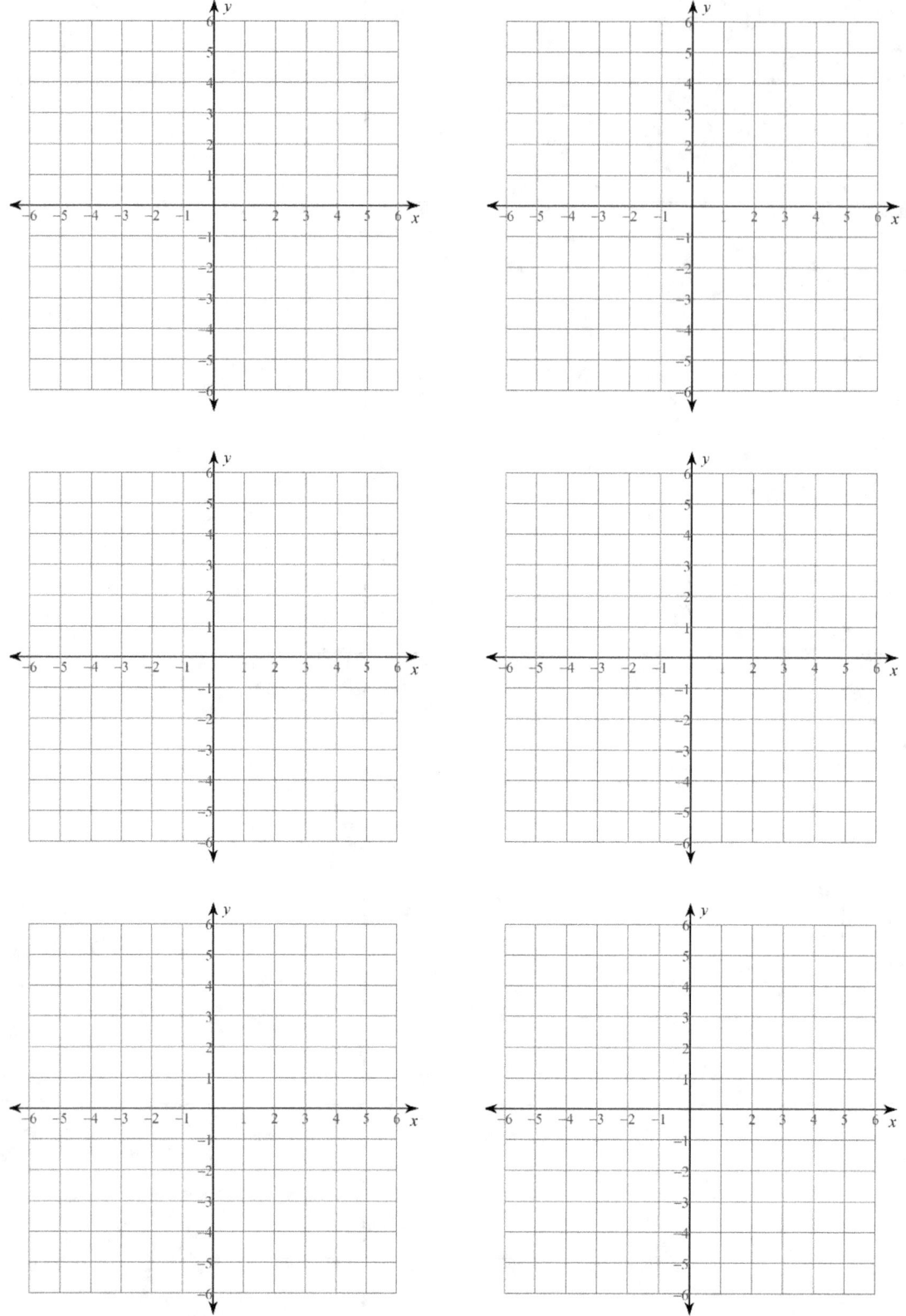

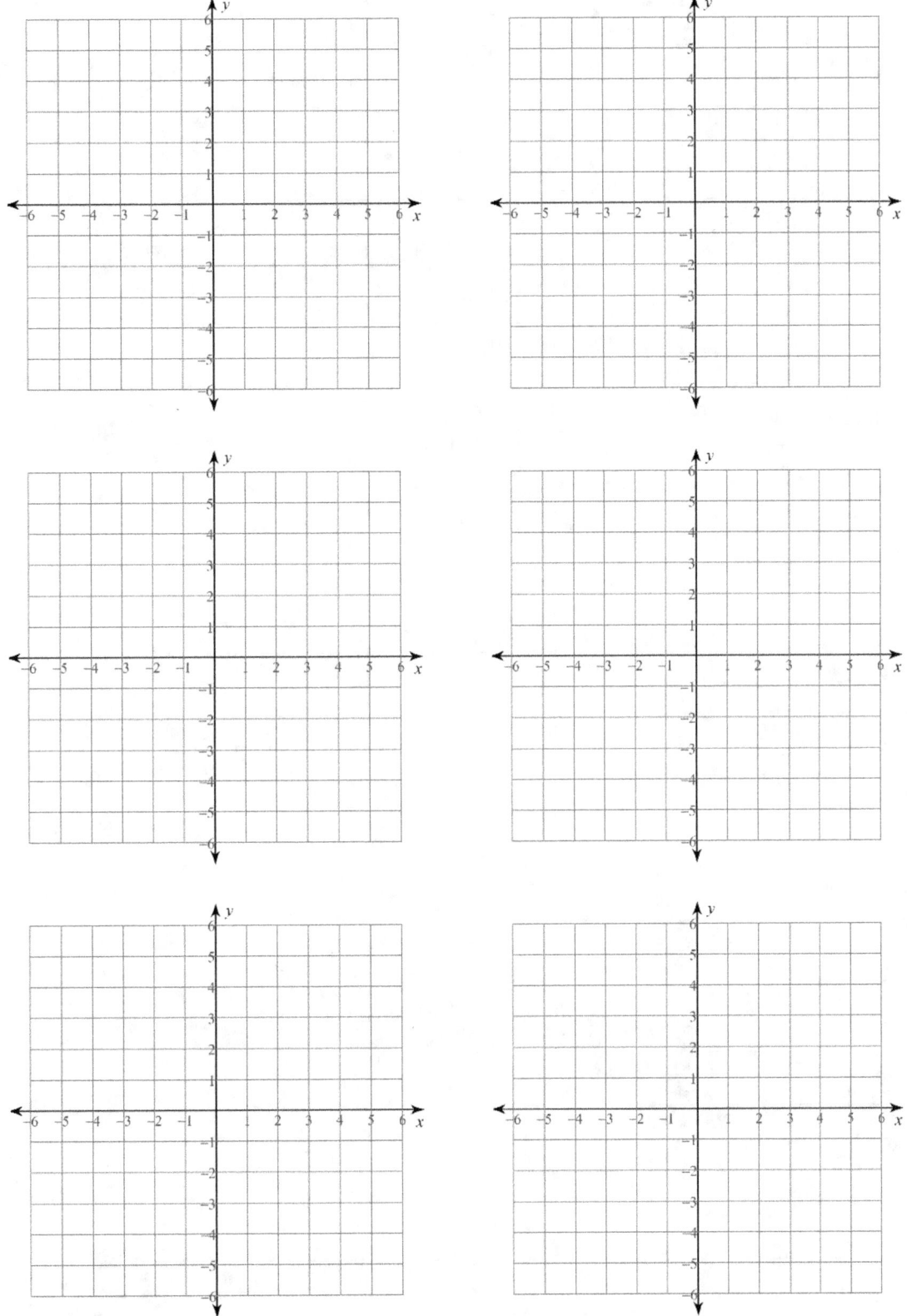

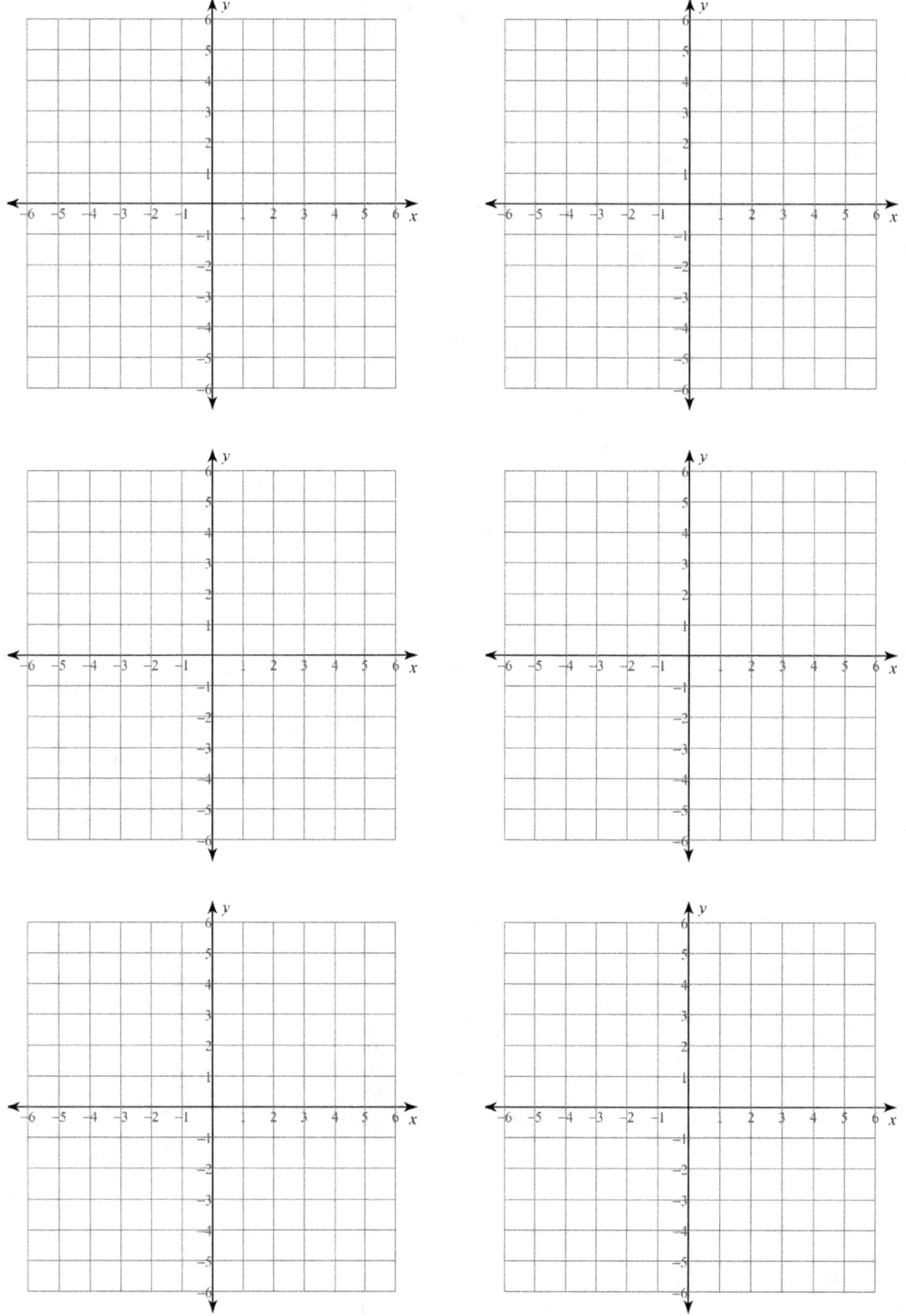

GRAPH PAPER TRIG

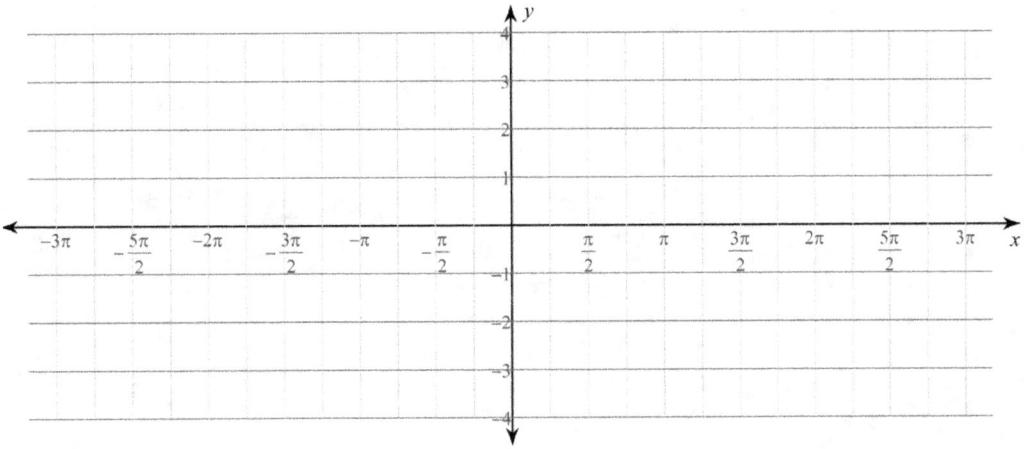

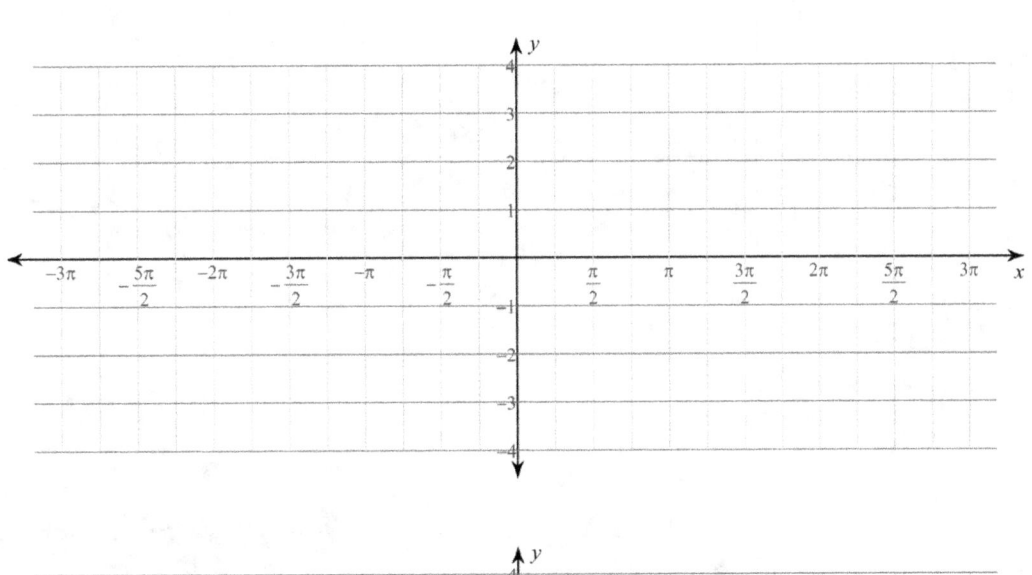

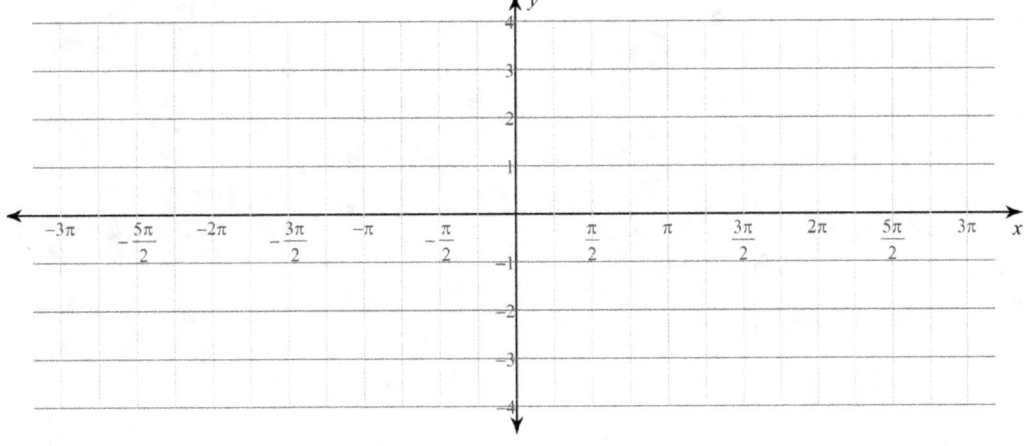

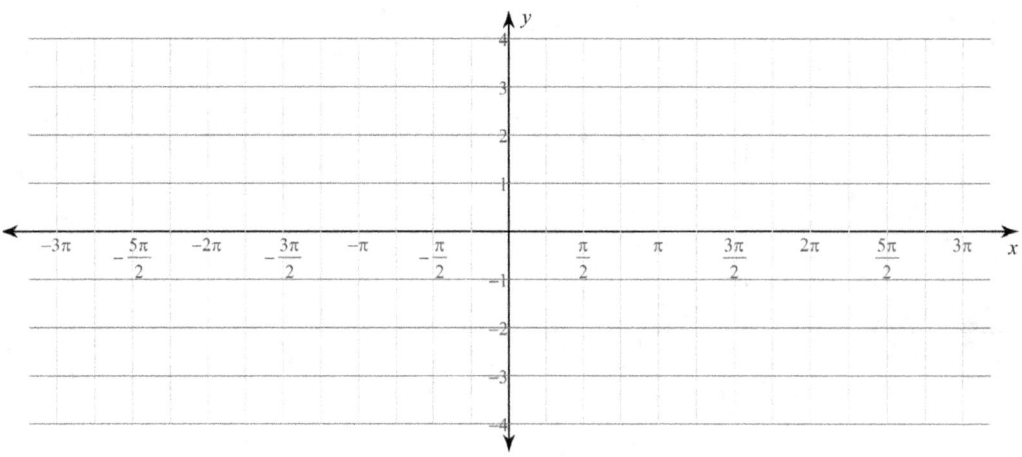

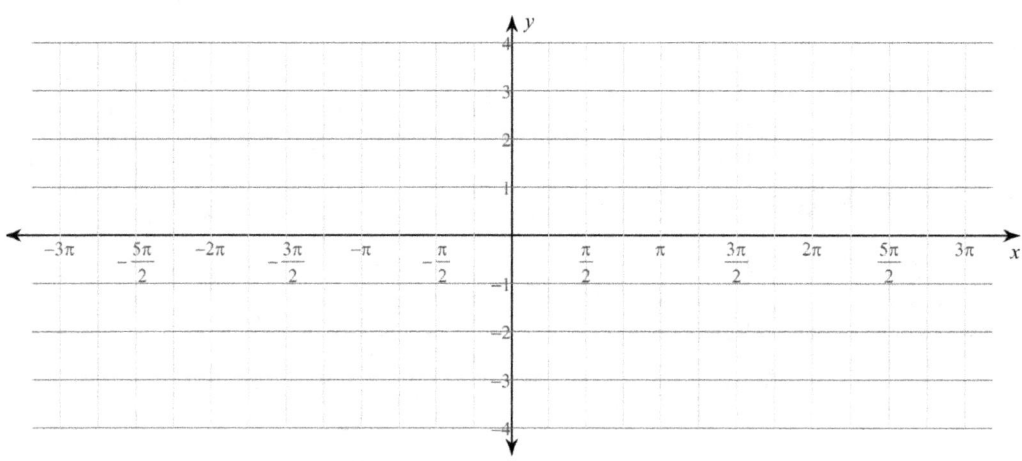

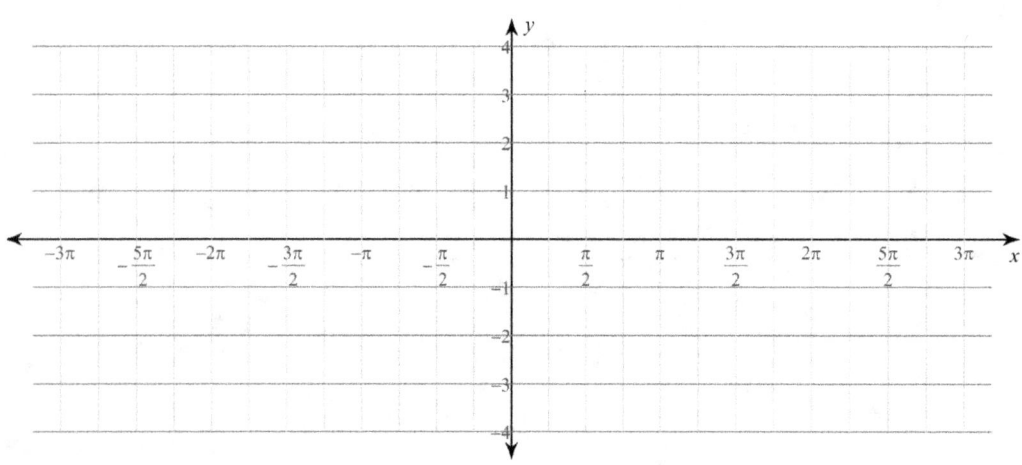

GRAPH PAPER FULL

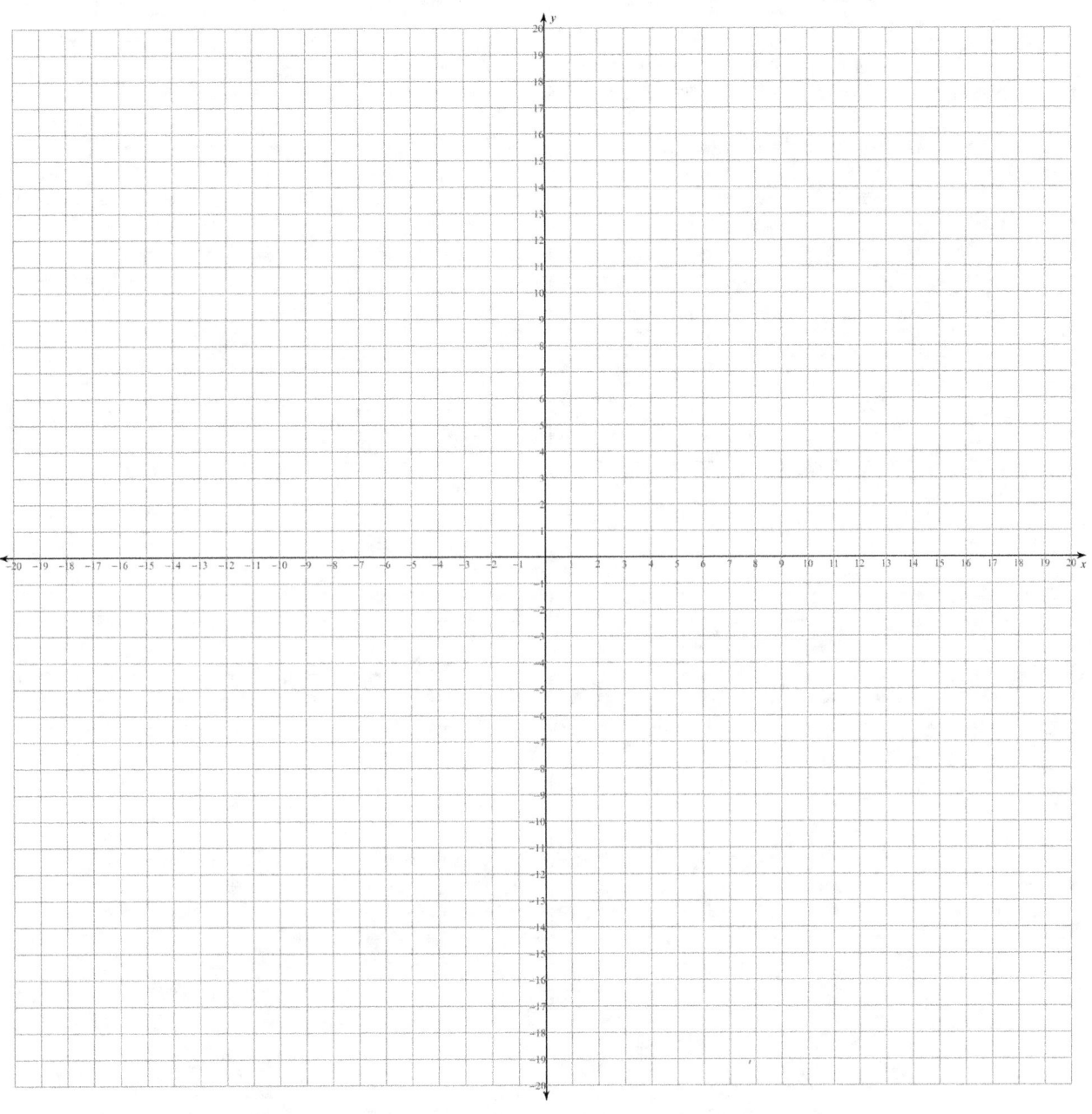

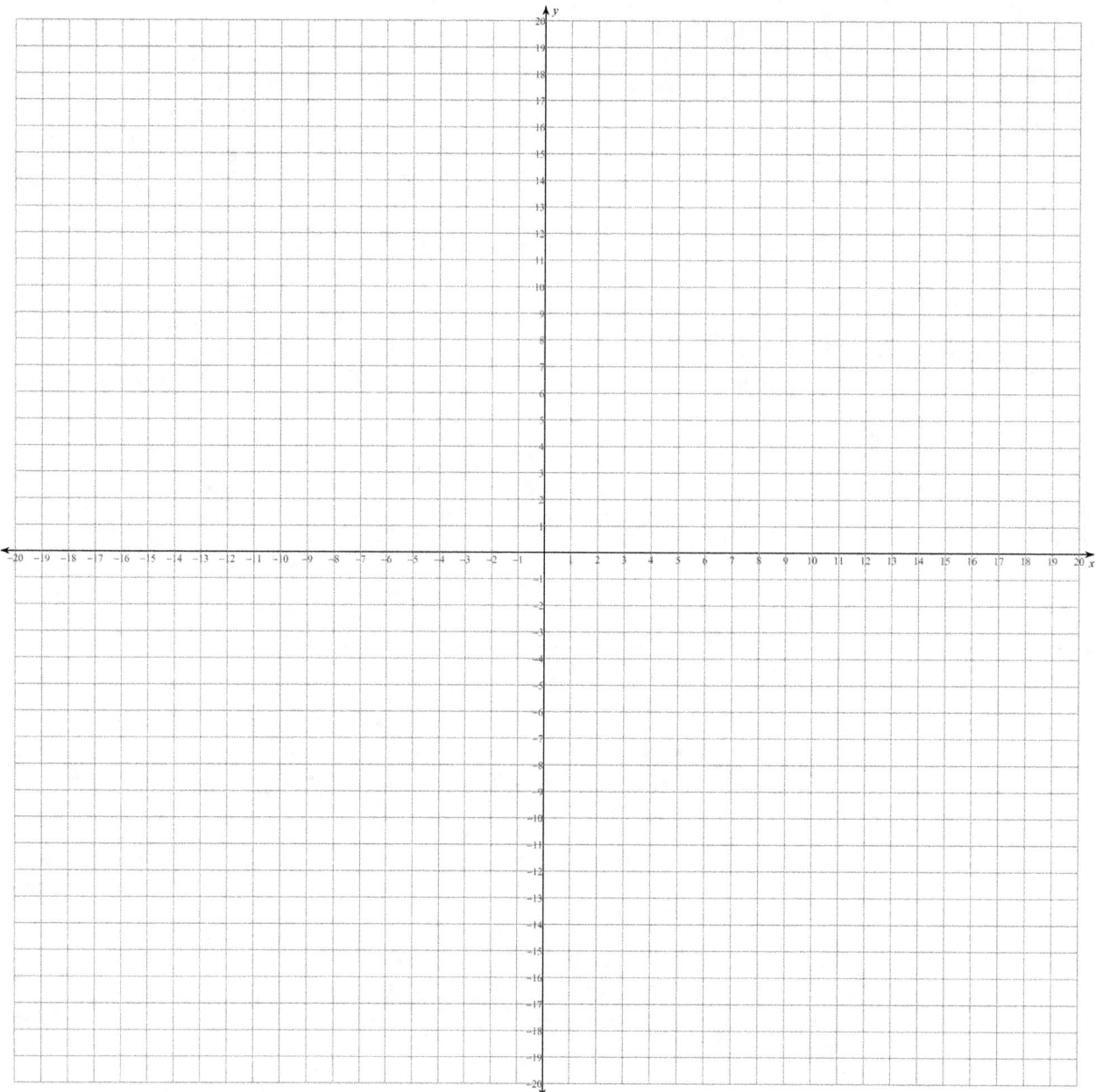

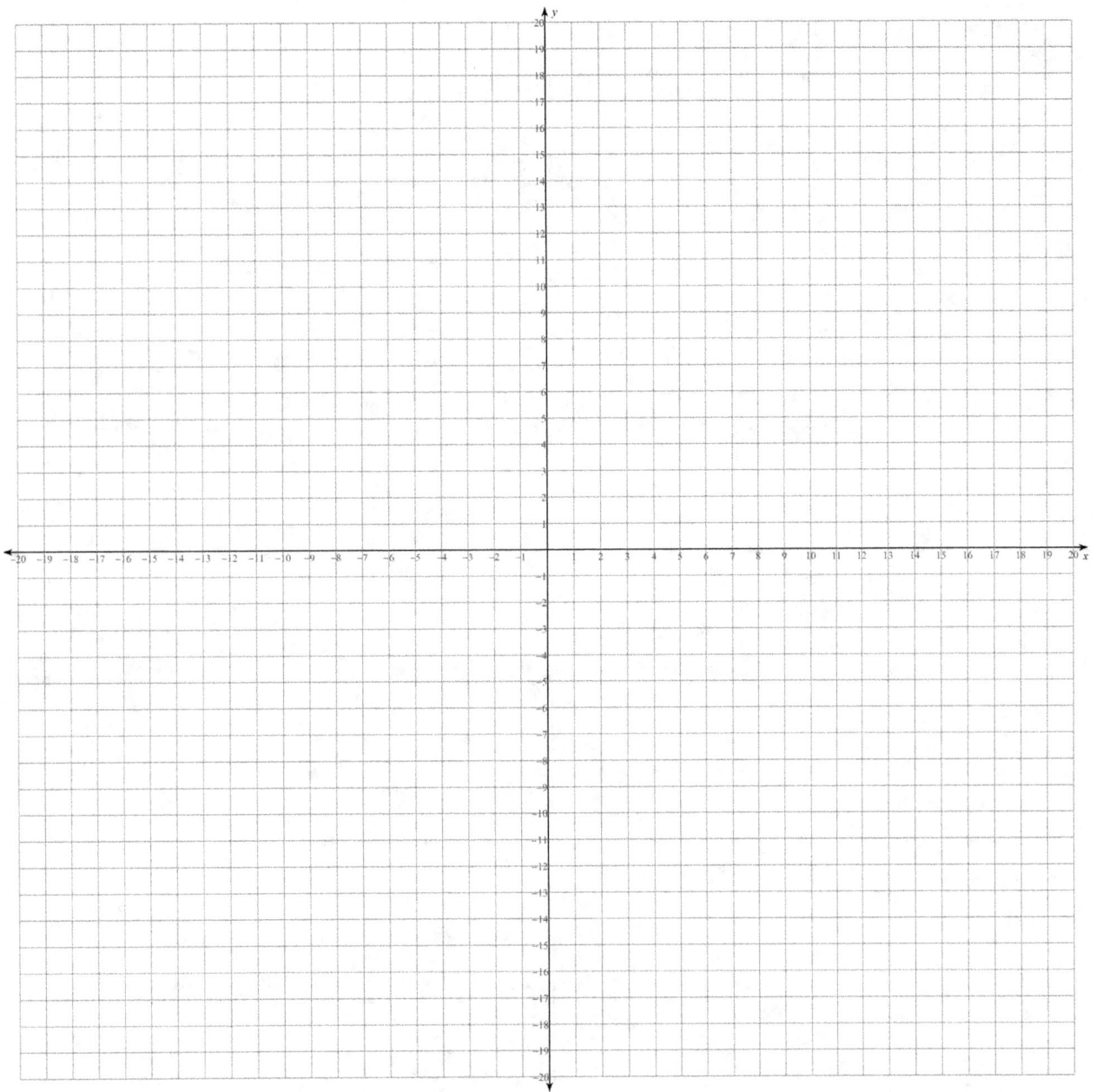

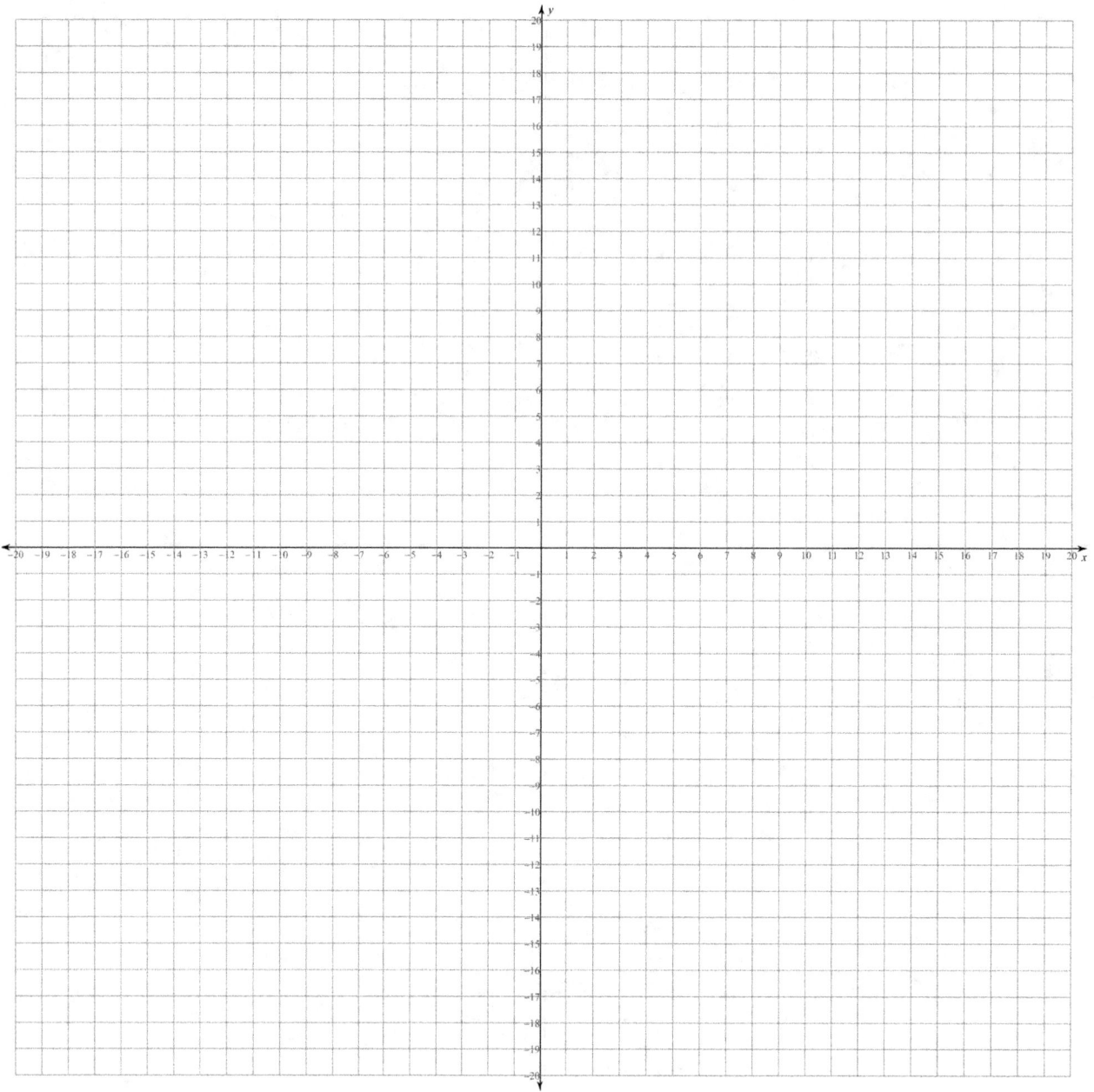

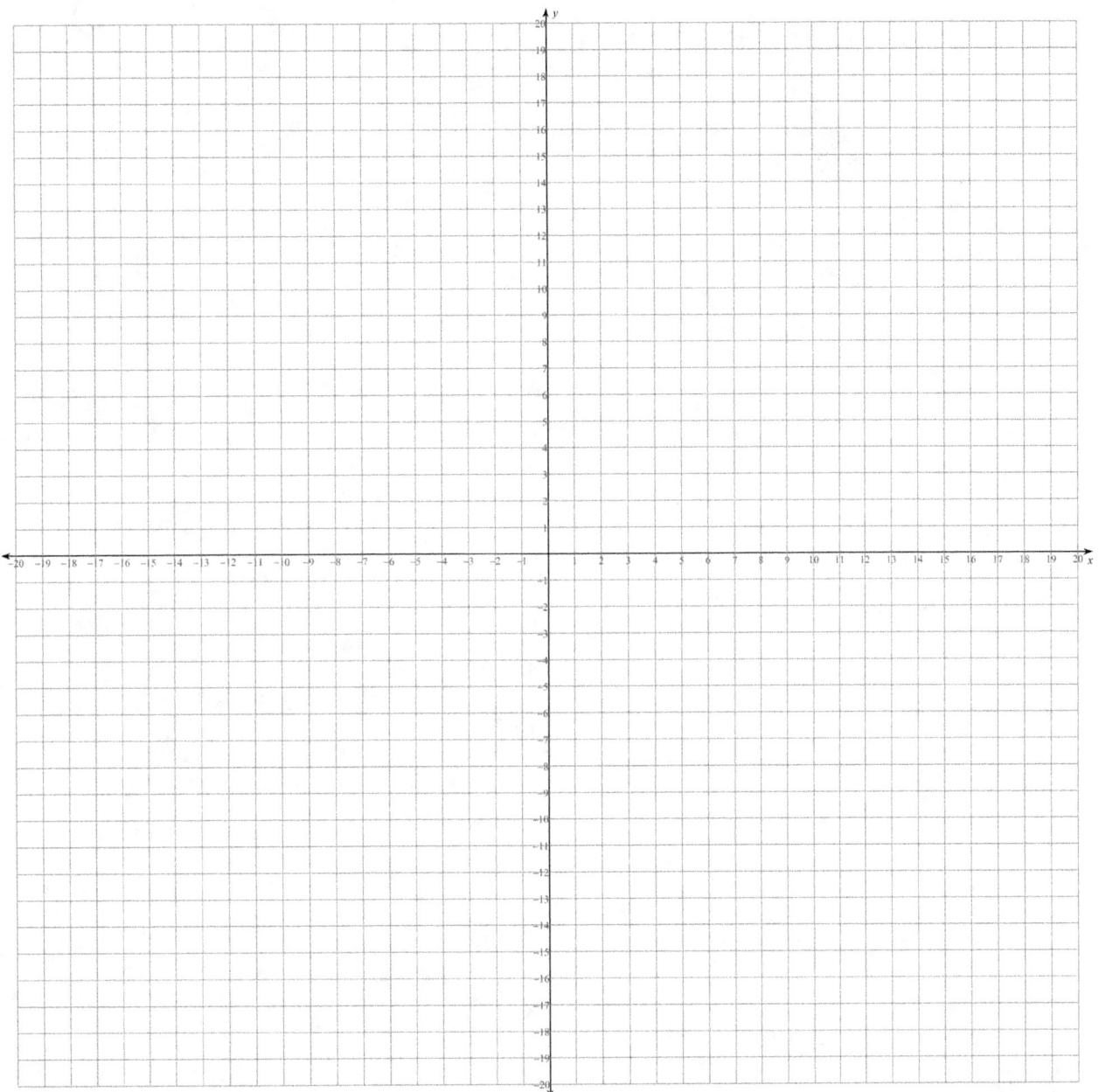

www.ingramcontent.com/pod-product-compliance
Lightning Source LLC
Chambersburg PA
CBHW060410220526
45465CB00008B/2826